Dr. D. K. Edwards
Professor and Chairman
Energy and Kinetics Department
School of Engineering and Applied Science
University of California, Los Angeles.

SOLAR COLLECTOR DESIGN

Revised Edition

THE FRANKLIN INSTITUTE PRESS℠

Current printing (last digit):
5 4 3
THE FRANKLIN INSTITUTE PRESS NUMBER: C/2-0078
ISBN Number: 0-89168-000-4
Library of Congress Catalog Card Number: 78-24684
Printed in the United States of America.

THE FRANKLIN INSTITUTE PRESS℠

Also in the Solar Energy Series:

Solar Cooling
Paul J. Wilbur and Susumu Karaki

Bioconversion: Fuels from Biomass
E. E. Robertson

Wind Energy
B. Wolff and H. Meyer

Thermal Storage
T. S. Dean

High Temperature Thermal Storage
R. Turner

Winds and Wind System Performance
C. G. Justus

ACKNOWLEDGMENTS

The author acknowledges helpful discussions with his colleague Professor Harry Buchberg and with Mr. Borek Vokach-Brodsky. Past grants from the National Science Foundation and a current one from the Energy Research and Development Administration enabled the author to perform research related to solar energy applications over the years from 1959 to present.

TABLE OF CONTENTS

LIST OF TABLES

1. INTRODUCTION

INTRODUCTORY REMARKS

Use of Solar Collectors. Solar heat collectors act simply to transform solar energy into heat energy. The heat can be used in the usual ways, to heat water for use in washing, to provide heating for winter comfort, to stimulate the growth of plants, to dry cereal grains and animal fodder, to drive heat engines for water pumping or power generation, and to drive thermal refrigeration machines for air conditioning or food preservation.

Solar Collectors and Their Main Components. Solar collectors are quite simple in concept. They consist usually of no more than four main parts. (1) An absorber acts to absorb the solar energy and so convert it into heat. In its plainest form it is no more than a blackened plate exposed to the sun. (2) A cover, in its simplest form a piece of flat glass, is usually placed over the absorber to reduce heat losses or, equivalently, to permit higher collection temperatures for fixed heat losses. (3) A concentrator may be used to increase the solar energy per unit area incident upon the absorber. The purpose again is to permit higher collection temperatures. A searchlight mirror can be used as a concentrator, for example. Even a flat mirror or a white-painted plate can be used to provide some concentration. Such a primitive concentrator is called a "booster." (4) A means of transferring the heat collected from the absorber to a transfer fluid, a so-called working fluid, is properly called a heat exchanger. Tubing soldered to the absorber plate, for example, serves as the heat exchanger in a simple solar water heater. Water circulated through the inside of the tubing is the working fluid which carries the heat from the collector to the point of use. Although these four main components are rather crude in their simplest forms, they may be made rather elegant or elaborate to good advantage by a designer with an appreciation for the laws of physics and the principles of engineering.

Factors in Solar Collector Design. While the heat from a solar collector can be used in the conventional ways that heat is used, there are a number of factors that make application of solar heat collectors challenging. Foremost there is the intermittent nature of solar radiation. Not only is there the nightly disappearance of the sun and the seasonal variation of the length of the day, but there is the unpredictable effect of local meteorological conditions upon the availability of solar energy. The diurnal and seasonal motions of the sun likewise must be contended with. Unlike fossil fuel heaters, solar heaters cannot be turned on whenever desired. To introduce dependability or ready availability, an auxiliary heater is almost universally required even when thermal storage is employed for better utilization of the solar energy. Thus the solar collector is used currently to augment or stretch out supplies of fossil, nuclear, or biological fuel, and not to replace them.

Another factor is the small amount of solar energy incident daily upon each unit area exposed to the sun. Somewhat less than one kilowatt is incident upon each square meter for a few hours per day (one square meter is 10.764 square feet, and a kilowatt-hour is 3412 British thermal units). Perhaps 4 to 8 $kwhr/m^2$ day can be collected. For winter heating of a moderate-sized, conventionally-constructed, single-family residence, a few hundred kwhr/day of heat energy is required. Thus approximately 50 to 200 m^2 of solar collector would be necessary for such a residence. The first cost (including the labor of installation) of such a collector is approximately 50 to 200 dollars per square meter and a 100 m^2 collector would cost on the order of $5,000 to $20,000. The facts that solar energy, like carpeting, comes by the square meter and that the first cost of the collector per square meter is appreciable make solar collectors capital intensive. Significantly large sums of capital must be invested in return for the use of the nondepletable source of energy.

The capital-intensive nature of solar collectors impacts importantly upon solar collector design. For example, a solar collector without thermal storage which is to be used for crop drying only during the few weeks after harvest must be made simply and inexpensively indeed, if it is worthy of consideration, for not much energy per unit area can be collected in such a short time, and thus the cost per unit energy collected is likely to be too high unless the collector is quite inexpensive. For this reason the walls of the silo or the storage bin itself, which are not chargeable to the solar collector, must be used as the absorber and/or part of the heat exchanger; and only paint, plastic film, and other such inexpensive extras can be afforded. In contrast, a water heater used year around to heat domestic or industrial process water collects a much larger amount of solar energy annually, and a greater expenditure for the solar collector is warranted, although it is still incumbent upon the designer to be as economical as possible.

Objectives of this Monograph. In order to design a solar collector and be assured that it is a reasonably good one, it is necessary to be acquainted with what has been done in the past, to understand the pertinent laws of physics which must be used to describe quantitatively how the device performs, and to know how to evaluate the relative merits of alternative designs. Such knowledge can be transmitted, and it is the objective of this work to introduce the reader to the requisite knowledge. Of course, once this knowledge is mastered, human ingenuity and art must be exercised to conceive designs whose merits can be weighed, and further human creativity is involved in the refinement of the most promising designs to achieve an optimum one. The author has no means of transmitting such creativity, but has to rely upon that with which each reader is endowed. Fortunately, the human race is well endowed with such creativity, and exercise develops it.

CLASSIFICATION OF DESCRIPTION OF SOLAR COLLECTORS

Working Fluid. Solar collectors can be classified by the working fluid used. Air heaters constitute one class and water heaters another. Of course, these two fluids are not the only ones used. Hydrocarbons or halogenated hydrocarbons are used, particularly when power generation or refrigeration is the goal, and antifreeze such as ethylene glycol is added to water, but the solar air heater typifies gas heaters, and solar water heaters typify liquid heaters, including those liquids which boil during the heating process.

When heat transfer fluids are rated by technical performance, those with a high thermal conductivity, a low viscosity, and a high specific-heat-density product rank high. On this basis air is least desirable, water intermediate, and liquid metals most desirable. Yet no liquid metal solar collectors are presently built, and some air heaters are. Other factors such as a low freezing point, noncorrosiveness, safety of life and property, and low cost overturn a simplistic ranking on the basis of heat transfer performance. When warm air is the desired goal as for drying or space conditioning, the noncorrosive nature of air permits the use of very inexpensive materials, and the prospect of a leak developing inadvertently poses no threat to property. Of course, when hot water is the goal, water itself or a water-antifreeze mixture is used in a unit made of copper or other corrosion-resistant materials.

Water Heaters. Two types of solar water heater systems may be distinguished, low pressure ones and high pressure ones. Low pressure systems are used for swimming pool heating, industrial water heating where recirculation is used or a pump to repressurize is not a significant factor, and domestic water heating in some parts of the world. Because of the low pressure, thin plastic film can be used to contain the water. The plastic must be thick enough to weather well but not so thick, when it is used as the absorber, that it interferes with the heat transfer, because plastic is a very poor thermal conductor. Also because of its low thermal conductivity, the plastic cannot be used as a fin; virtually the entire absorber surface must be wetted by the water. Figure 1–1 illustrates a simple low-pressure plastic water heater.

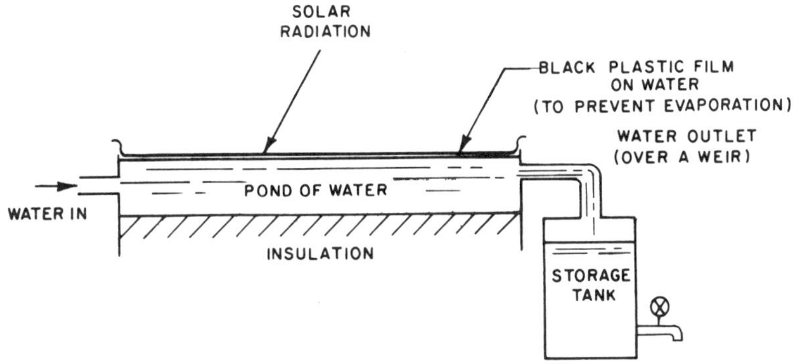

1-1. A Simple Low-Pressure Water Heater

In a high pressure water heater the water is usually contained in copper tubes, and metal plates called fins are used between the tubes to give additional absorber area. Figure 1–2 shows a conventional solar water heater. Steel, aluminum, or copper may be used for the fins. Because of its high thermal conductivity, a fourth of a millimeter of copper will do for the fins; a half of a millimeter of aluminum and two millimeters of steel are necessary because of their lower thermal conductivities. The spacing between tubes is affected by the choice of fin thickness as will be shown when the topic is taken up in detail.

Air Heaters. Heat transfer coefficients in air are approximately two orders of magnitude lower than in water. Thus while the wall-water surface area is of minor concern in the design of solar collectors without high concentration, it is a concern for air heaters, even when no concentrator is used. If the area available for heat exchange is no greater than the projected area of the absorber, that is, the area projected onto a plane perpendicular to the sun's rays, then the absorber temperature may have to be 10 to 50° C (18 to 90° F) greater than the air itself to transfer the solar energy to the working fluid. If the absorber is unnecessarily hot, the heat transferred by it to the cover glass is unnecessarily high, and some energy is lost. A ratio of heat exchange area to projected absorber area between two and ten is thus quite desirable, and means of obtaining such a ratio have been dominant features in air heaters.

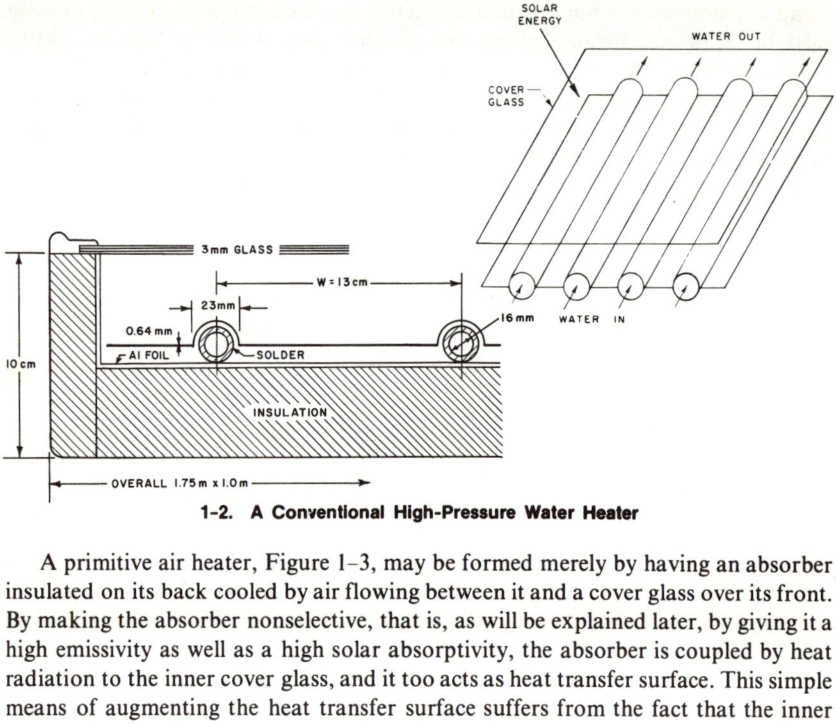

1-2. A Conventional High-Pressure Water Heater

A primitive air heater, Figure 1-3, may be formed merely by having an absorber insulated on its back cooled by air flowing between it and a cover glass over its front. By making the absorber nonselective, that is, as will be explained later, by giving it a high emissivity as well as a high solar absorptivity, the absorber is coupled by heat radiation to the inner cover glass, and it too acts as heat transfer surface. This simple means of augmenting the heat transfer surface suffers from the fact that the inner cover glass does not work well in preventing heat losses from the absorber and working fluid.

To use the inner glazing as a guard to reduce heat losses, it is necessary to transfer heat to the air from the *back* of the absorber. The absorber then serves as the front of the air duct, and the inner lining of the insulation serves as the back of the duct. By giving the back of the absorber and the back of the duct high emissivities, some augmentation of heat transfer surface occurs, but by corrugating the absorber as shown in Figure 1-4, still further augmentation is obtained. Corrugations in the form of 60° vee grooves give twice as much heat exchange surface on the back as projected area on the front. The back of the duct gives additional area, and, if it is corrugated, still more. The benefit obtained from the back of the duct depends upon the convective heat transfer coefficient in the duct.

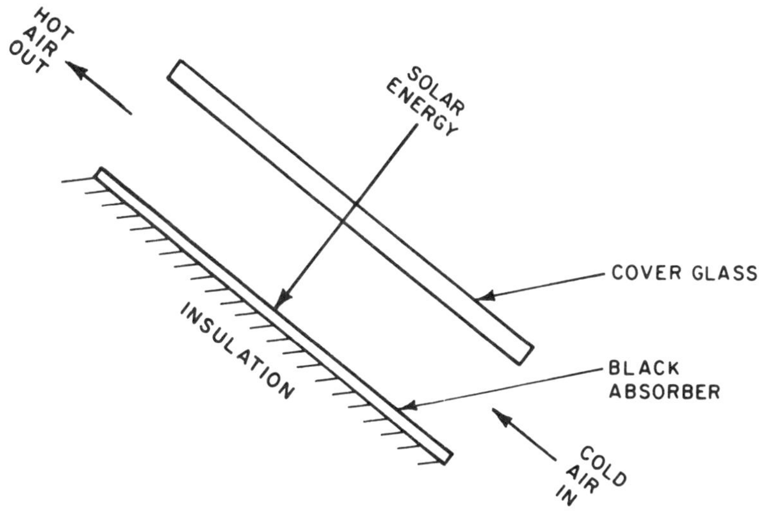

1-3. A Primitive Air Heater

Another method of obtaining a high ratio of heat transfer surface to projected absorber area is to employ a porous absorber. Black cloth, blackened fiberglass, slit and expanded screens, wire mesh, or slats can be used as shown in Figure 1-5.

Orientation. Solar collectors can also be classified as fixed or tracking. A fixed collector is one which does not move during the course of a day. The usual domestic water or space heater is fixed. In the Northern Hemisphere they face south and are tilted from the horizontal at an angle equal to the latitude, to favor spring and fall or year around collection. A 20° greater tilt or even a vertical orientation favors winter collection, while a 20° lesser tilt or even a horizontal orientation favors summer collection. Such fixed collectors have the advantages of simplicity, reliability, and low maintenance, and they may be built into a structure, that is, they may serve as integral parts of a roof or wall so that their cost need not be fully charged to the heat collected. On the other hand, fixed collectors are often irradiated by the sun at unfavorable angles, that is, the sun may strike the collector at grazing angles so that not much of a shadow is formed; the solar energy intercepted is proportional to the area of the shadow normal to the sun's rays. A mount which allows the collector to always face the sun exposes the area to maximum irradiation. Whenever high concentration is required to achieve high temperatures, the collector (or some component of it) must track the sun rather accurately.

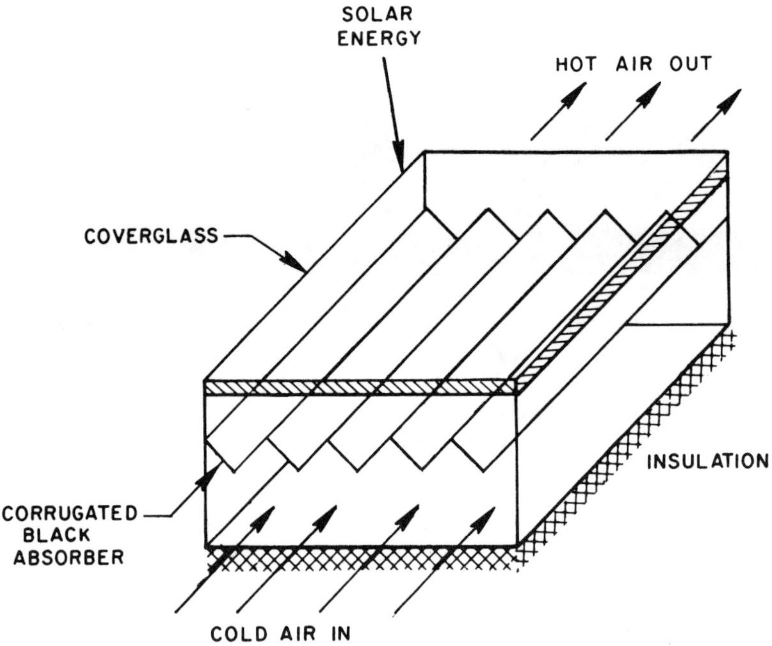

1-4. A Corrugated-Absorber Air Heater

Semifixed collectors may be used. For example, a booster may change position once a day, or its angle may be adjusted seasonally. A cylindrical mirror aligned east-west may be fixed during the course of a day or week but the mirror or the absorber may be positioned periodically to adjust to the change in the noontime angle of the sun through the course of a year. Even a simple flat plate collector may be tilted to near vertical in the winter and near horizontal in the summer to obtain more favorable angles for solar irradiation.

Type of Component. The type of cover, absorber, heat exchanger, or concentrator is often used to describe a class of solar collectors. The cover may be nonexistent, in which case the collector is referred to as unglazed, or one or two glass panes may be used, in which case the appellation single-glazed collector or double-glazed collector may be applied. A honeycomb solar collector is one which employs a honeycomb cover to reduce infrared radiation and/or natural convection heat losses. A honeycomb-like cellular structure is located between the absorber plate and outer glazing. Solar radiation passes down the passages being reflected and/or transmitted toward the absorber. Infrared radiation from the absorber is intercepted by the sides of the honeycomb, which should be absorbing to infrared radiation, and the radiative transfer is thereby reduced. Natural convection, buoyant movement of air from the hot absorber plate to the cold cover glass and vice versa, is suppressed by the honeycomb walls. In this way heat losses by both radiation and air movement are reduced. The type of honeycomb may be described also in classifying solar heaters. Adjectives describing the material; glass, plastic, or paperboard; may be used. The shape, hexagonal or rectangular, may be mentioned. Transpired or nontranspired honeycomb refers to whether or not air is drawn downward through the honeycomb. A transpired honeycomb collector shown in Figure 1-6 makes a good solar air heater. Cylindrical evacuated covers may also be used, Figure 1-7, to eliminate natural convection between the absorber and cover.

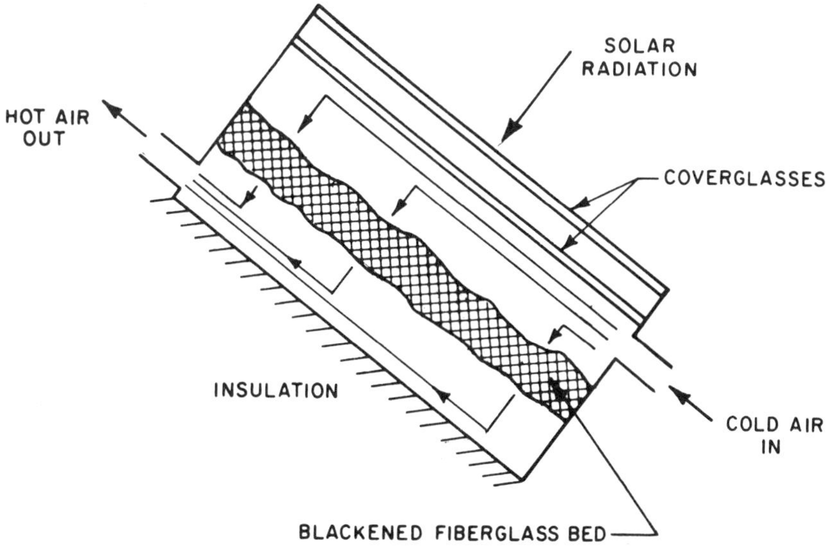

1-5. A Porous-Absorber Air Heater

Absorbers may be classified by shape, for example as flat plate or cavity absorbers. Their radiation characteristics are also described as selective or non-selective. A selective absorber is one with a specially treated surface which acts as a black surface absorbing the radiation for the energetic photons of short wavelength in the solar spectrum but behaves as a metallic mirror reflecting infrared radiation of longer wavelengths

The type of heat exchanger is also used to classify solar collectors. A tube-sheet collector, for example, is one with tubular heat exchanger passages integral with or soldered to a flat plate sheet collector. A porous-bed air heater is one which uses, for example, blackened fiberglass wool as the absorber and heat exchanger. The vee-groove air heater has been previously mentioned.

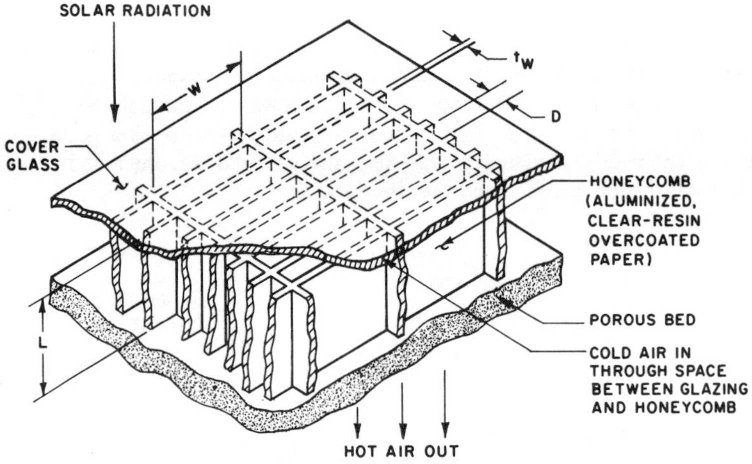

1-6. A Transpired Honeycomb Solar Collector

A variety of concentrators have been used for solar collectors, and the type employed is often used to classify the collector. Common are the paraboloid, like a searchlight mirror, and the cylindrical parabola, a long mirror only singly curved. A field of mirrors each mirror tracking the sun and focusing the solar radiation upon an absorber mounted upon a tower or similar structure has been used for solar furnaces and proposed the solar heat power generation. Figure 1–8 depicts such a system.

PERFORMANCE OF SOLAR COLLECTORS

Collector Efficiency. One measure of technical performance is the instantaneous collector efficiency. It is defined as the ratio of the heat power collected by the working fluid divided by the heat power incident from the sun. This latter value, the denominator, is the area of the collector projected normal to the sun's rays multiplied by the solar irradiation. The former quantity is the net heat collected, that is, that remaining after the losses from the absorber.

The collector efficiency achieved by a given unit depends primarily upon the level of the solar irradiation and the difference in temperature between the absorber and the ambient air. When the absorber temperature is high, the heat losses from it are large, and the collector efficiency is low. On the other hand, if, as may rarely occur with swimming pool heaters, the absorber temperature is 10° C (18° F) or so *below* the ambient temperature, a simple unglazed flat plate collector may have an efficiency approaching or even exceeding unity. When such a primitive collector operates 30° C (54° F) above ambient, its efficiency may drop to zero. For a more intricate flat plate collector; for example, a double-glazed selective black unit, a honeycomb nonselective black collector, or an evacuated cylindrical glass unit with selective absorber; the collector efficiency can exceed 30 percent with collection temperatures above 100° C (212° F). With a concentrator much higher collection temperatures may be achieved with comparable efficiency.

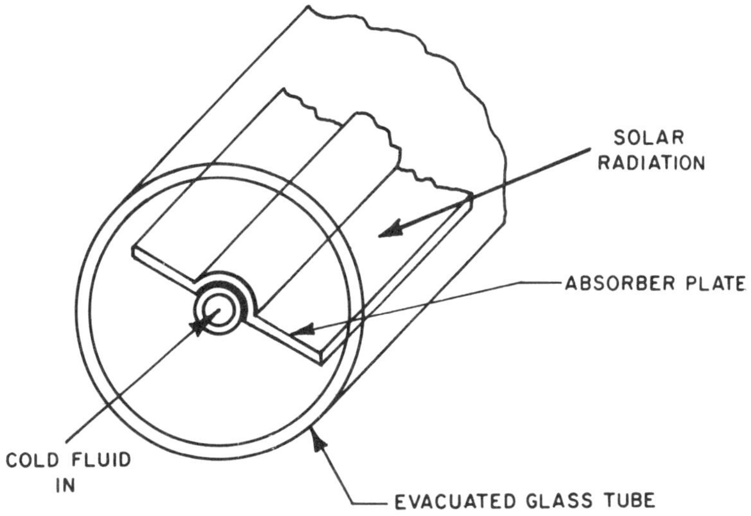

1-7. An Evacuated Glass Cover Collector

Collector efficiency can also be averaged over the course of a day, a season, or a year. The meaning is the ratio of the useful energy collected to the solar energy intercepted during the time period.

Cost of Solar Energy as a Design Criterion. The variety of solar collector types and the variable parameters that must be fixed (such as the angle of orientation, the temperature of collection, the spacing of the collector tubes, and the number of cover glasses) require the solar collector designer to make a large number of decisions. An aid which helps the designer choose good alternatives is the estimation of the cost of the energy collected. If all costs are correctly allowed for; the costs of the materials and labor, the maintenance, taxes, and insurance; the designer chooses a system with the lowest cost consistent with his or her sense of aesthetics.

There are, of course, considerable difficulties in estimating or assigning costs. Maintenance, for example, is often really remedial work to repair the results of faulty design or poor workmanship. Almost by definition the designer cannot foresee the results of faulty design. Because long, continuous experience with solar collectors is lacking, it is difficult to make adequate but not overly costly provision for ground or foundation movement, cyclical thermal expansion and contraction, photo-degradation, abrasion, and corrosion, the actions of wild or domestic animals, including man, wind forces and sonic booms, and falling hail stones, fruit, or tree branches.

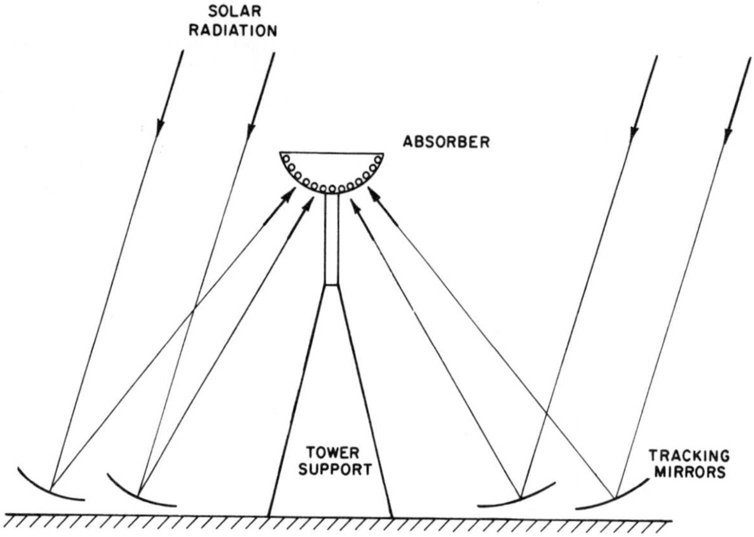

1-8. A Field of Mirrors Concentrator

Another difficulty at arriving at proper solar energy cost assignments is the proper allowance for the fact that the energy is received in the future while a large portion of the costs are borne initially. What is done is to regard the initial costs as a capital investment that ties up resources which could have been invested elsewhere for a certain percent per annum return and which in addition must be recovered by assigning an additional cost to the solar energy collected. The combined cost for the use and recovery of the initial investment is an annual charge to the ownership and operation of the solar collector system. To this annual charge is added whatever annual maintenance, taxes, and insurance charges there may be. The total annual cost is then divided by the total annual useful energy collected to obtain a cost per unit of energy, so many cents per kilowatt hour.

While the difficulties in forecasting useful lifetime, in anticipating maintenance, and in assigning appropriate interest rates are indeed great, one should make the effort to do so, because one does gain a useful design criterion. Particularly when comparing two types of solar collectors, one finds that many of the uncertainties apply almost equally to both systems being compared and so cancel in large part, and the system with the least cost charged to the energy collected really is the better of the two.

Perhaps the lesson of greatest benefit to the designer in estimating the cost of solar energy is the discovery that somewhat undersized solar collectors make the best use of available resources. By designing the unit to accommodate minimum needs, the designer ensures that all or nearly all the solar energy collected is put to use and none is wasted.

2. HEAT LOSS CONTROL

LOSS MECHANISMS

The solar heat energy available from a solar collector is that incident less that lost. Thus the designer orients the collector so as to maximize that incident and takes steps to minimize that lost. Heat energy is lost by reflection, reradiation, convection, and conduction as portrayed in Figure 2-1.

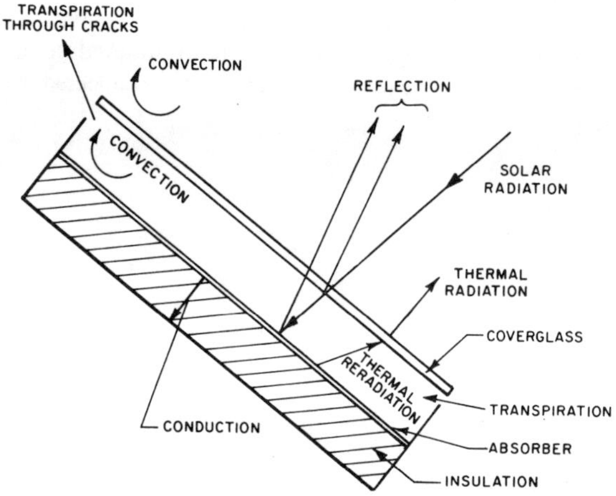

2-1. Heat Loss Mechanisms

Reflection occurs from the cover glass, if any, and the absorber itself. If no reflection occurred, the collector would look perfectly black when viewed from the direction at which the sun is incident. An ordinary glass reflects 7 to 8 percent of the incident radiation, and even black paint reflects 2 to 6 percent. Scattering from scratches or dust on mirrors or covers is typically 0.25 to 1.5 percent or so per interface. Scattering from a salt-encrusted solar still or from a crazed plastic cover may be many times larger. Misalignment or distortion of mirrors may result in reflected radiation missing the absorber.

Reradiation occurs from hot elements of the solar collector. The absorber is hot and radiates heat to space or the cover glass. An infrared-opaque cover glass when present serves to reduce the net radiation loss by absorbing the radiation from the absorber and reradiating approximately half of it back to the absorber. The remainder is reradiated to space and lost.

Convection is the transfer of heat by conduction into a moving fluid, the atmospheric air. A cover glass slows the motion of the air over the absorber and so reduces the convective heat loss. However, in the space between the absorber and cover glass, hot air tends to rise and cool air fall thus creating a buoyancy-driven convection, termed natural convection. If the space between the cover and absorber is not well sealed, leakage of air into and out of the space can occur; this leakage is called transpiration. It arises from a chimney-like effect, often augmented by wind-induced pressure gradients. Hot air leaves the space while cold air is drawn in to replace it. Natural convection occurs also on the outside of the cover glass, or, if the wind is blowing, forced convection occurs. The cooling of the cover glass increases the heat loss from the absorber to the cover glass.

In a solar still, which is used to distill fresh water from salty water, any transpiration that carries away moisture is particularly insidious. The water vapor carries away the large latent heat of vaporization.

Conduction is the transfer of heat within a material having a temperature gradient. It occurs from the microscopic motions of the molecules, atoms, or electrons within the material. In order to reduce conduction a poor thermal conductor, insulation, is located on the back of a flat plate collector. What heat is conducted away from the absorber through the insulation or through supports is eventually lost by radiation and convection.

Concentration by mirrors serves to reduce the *fractional* heat losses by reradiation, convection, and conduction, because for the same absorber area a much greater collector area is deployed. Only a small amount of the solar radiation is absorbed by the mirror collector. This small amount is nearly all lost by radiation, convection, and conduction, of course.

ANTIREFLECTION COATINGS

The reflectance r of a smooth plane interface between air or vacuum and a material is, for normal incidence,

$$r = \frac{(n-1)^2+k^2}{(n+1)^2+k^2} \tag{2-1}$$

where r is the reflectance of the interface, n is the refractive index, and k is the absorptive index ($\underset{\sim}{n} = n-ik$). (The quantity r as used here is not to be confused with the Fresnel coefficient.) This equation shows that the lower k, the lower the reflection, and the closer n is to 1 the lower the reflection. Ideally one would like to use as a cover glass a material with $k \simeq 0$ and $n \simeq 1$, but no such material exists. Glass, which is readily available and inexpensive, has $n \simeq 1.5$. It is usually used as is, and the reflection losses accepted. Consequently the value of r is 0.04.

Glass has a thickness d usually a few millimeters, while the wavelength λ of the solar irradiation varies from approximately 0.3 to 3 micrometers. Thus the glass is 1000 to 10000 times thicker than the wavelength of the solar radiation. Slight variations in the glass thickness and averaging over the solar wavelengths eliminates interference effects so that the reflectance and transmittance of the glass pane are respectively

$$\rho = r + \frac{r(1-r)^2t^2}{1-r^2t^2} \tag{2–2}$$

$$\tau = \frac{(1-r)^2t}{1-r^2t^2} \tag{2–3}$$

where the quantity t represents the transmittance within the glass

$$t = e^{-4\pi k\delta/\lambda} \doteq 1 - 4\pi k\delta/\lambda \tag{2–4}$$

In addition to wanting a low value of pane reflectance ρ and hence a low value or surface reflectance r, one wants a high value of pane transmittance τ and hence a high value of glass transmittance t. It is thus important to have a small value of absorptive index k. An effective value of k/λ can often be employed for the solar spectrum; a commercial window glass tested by the author had $k/\lambda \doteq 1.3 \times 10^{-6}\ \mu m^{-1}$. For such a glass a piece 3 mm thick has a value of $t = 0.952$, $\tau = 0.879$, and $\rho = 0.0735$. Glass with low iron content has a desirably low value of k.

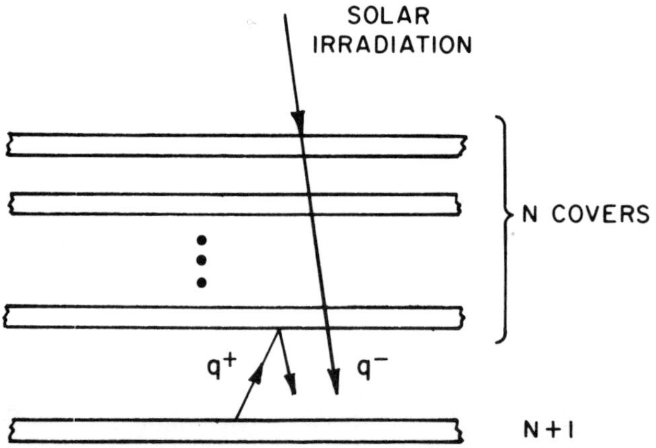

2-2. Embedding One More Cover Glass

A reflection loss ρ of 7.35 percent occurs with the use of one cover glass, and a greater loss occurs when multiple cover glasses are used. The reflectance and transmittance of a system of multiple covers may be found readily by an embedding calculation. Let one more cover glass be added under a stack containing N covers as pictured in Figure 2-2. The N-stack has system reflectance ρ_N and transmittance τ_N. Let a solar flux of 1 be incident on the top of the N stack. Then the downward flux q^- from the N stack is

$$q^- = \tau_N + q^+\rho_N \qquad (2\text{–}5)$$

where the upward flux q^+ is simply

$$q^+ = \rho q^- \qquad (2\text{–}6)$$

Solving gives

$$q^- = \frac{\tau_N}{1-\rho\rho_N}, \quad q^+ = \frac{\rho\tau_N}{1-\rho\rho_N} \qquad (2\text{–}7)$$

The flux transmitted through the added cover is τq^-, and this flux, since that incident is unity, is the new system transmittance τ_{N+1}

$$\tau_{N+1} = \frac{\tau\tau_N}{1-\rho\rho_N} \qquad (2\text{–}8)$$

The new system reflectance ρ_{N+1} is the former one ρ_N plus the transmitted upward flux $\tau_N q^+$,

$$\rho_{N+1} = \rho_N + \frac{\rho\tau_N^2}{1-\rho\rho_N} \qquad (2\text{–}9)$$

Note in passing that as N goes to infinity and τ_N goes to zero, adding one more glass leaves ρ_{N+1} invariant from ρ_N.

One commences with values of $\rho_1 = \rho$ and $\tau_1 = \tau$ for a single cover and obtains ρ_2 and τ_2 and then ρ_3 and τ_3 etc. Table 2–1 shows values for $\rho = 0.0735$ and $\tau = 0.8790$. To illustrate the effect of antireflection coatings on low-iron glass, values are also shown for $\rho = 0.040$ and $\tau = 0.950$.

The table shows clearly the advantage of a low-reflecting, low-absorbing glass when multiple covers are desired. For example, when three covers are used, the good glass gives 25 percent more transmission to the absorber. When account is taken of the other heat losses, the 25 percent improvement in transmission may lead to a 50 percent or more improvement in collection efficiency. Also, the reflection from multiple covers is seen to be appreciable, and this reflection may be troublesome to the architect.

How does one obtain a desirably low value of r? One of two stratagems can be employed. The first alternative because of its expense is economical at present only when concentrators are used. This alternative is to apply optical antireflection coatings in the form of multiple thin films to the glass surfaces. The other stratagem is to chemically microetch the glass surfaces.

TABLE 2-1

REFLECTANCE AND TRANSMITTANCE OF A STACK OF COVER GLASSES

Number of Covers	A Commercial Glass		A Hypothetical Glass	
	System Reflectance	System Transmittance	System Reflectance	System Transmittance
0	0.0000	1.000	0.0000	1.000
1	0.0735	0.879	0.0400	0.950
2	0.1306	0.777	0.0716	0.904
3	0.1754	0.689	0.1089	0.861
4	0.2101	0.614	0.1388	0.822
5	0.2389	0.548	0.1659	0.785
6	0.2614	0.490	0.1907	0.751

The optical antireflection coating in its simplest form is a layer with thickness $\delta = \lambda/4n$ of a low index of refraction material on the surface. The reflected radiation from the glass-coating interface is out of phase with the radiation fro n the coating-air interface, and the waves cancel much as sound waves cancel when they beat. Two problems make a single coating rather ineffective. First a coating material with a desirably low value of n is not available, and, more importantly, reinforcement rather than cancellation occurs for the solar radiation with a somewhat shorter wavelength. As a consequence another pair of layers are used to obtain cancellation of the pair-reinforced wavelengths, and additional layers broaden the spectral band over which an antireflection effect is obtained. Very effective broad-spectral-band antireflection coatings result from eight or so layers, but the expense is presently a factor of ten or more too great for flat plate systems of no concentration.

Antireflection by microetching is not as simple to describe as that by thin films, which is described by the Fresnel relations. A crude model of the process suffices to give an understanding, however. By chemically dissolving a greater portion of the glass near the surface than at some distance below the surface, a desirable gradient in effective refractive index is created. The effective refractive index is reduced near the surface and climbs as the radiation proceeds into the glass. In effect a thin nonhomogeneous film is created on the surface of the glass. When the thickness of a nonhomogeneous film and the variation of its index of refraction are correct, a very effective antireflection coating results. The goal of microetching to obtain a solar antireflection film should not be confused with that of making picture glass matt or diffuse. In the former case the goal is not merely to render the reflection diffuse, as in the latter case, but to reduce the reflection regardless of whether it is specular (mirror-like) or diffuse.

To the author's knowledge reliably microetched glass is not available in commercial quantities at present. Note that Eq. (2–8) can be used to obtain the fraction of solar radiation absorbed by an opaque absorber plate merely by replacing transmittance τ with solar absorptance $\alpha = 1-r$ when the absorber is reached during the embedding process. Similarly Eq. (2–9) applies. These relations show the desirability of a good black low-r absorber.

CONTROL OF RERADIATION

Cover Glasses as Radiation Shielding. Reradiation may be reduced simply by using multiple cover glasses. As a simple illustration consider a black-painted hot absorber plate with an infrared emittance of ϵ_1. The absorber is imagined to be at temperature $T_1 = 80°C = 353.15°K$. An outside cover with infrared emittance ϵ_2 is imagined to be at $T_2 = 20°C = 293.15°K$. With no other cover glasses the heat flux lost from absorber to cover would be q, with units of w/m^2,

$$q = \frac{1}{\frac{1-\epsilon_1}{\epsilon_1} + \frac{1-\epsilon_2}{\epsilon_2} + 1}[\sigma T_1^4 - \sigma T_2^4] \qquad (2\text{–}10)$$

where $\sigma = 5.6697 \times 10^{-8}\ w/m^2K^4$ is the Stefan-Boltzmann constant. Typical values of ϵ_1 and ϵ_2 are 0.9 and 0.8 respectively. Hence q would be

$$q = \frac{1}{\frac{1}{0.9} + \frac{1}{0.8} - 1}[5.6697 \times 10^{-8}][353.15^4 - 293.15^4]$$

$$q = (0.7347)(463.1) = 340\ w/m^2$$

For a flat plate collector the solar irradiation might be 700 w/m^2, so that the above value of 340 w/m^2 would constitute a loss of nearly 50 percent of the solar energy. Of the solar energy reaching the absorber and absorbed by it, (700)(0.879)(0.95) = 585 w/m^2, the loss of 340 w/m^2 by heat radiation would be nearly 60 percent. Here a solar absorptance of 0.95 has been assumed for the absorber plate.

If N–1 additional cover glasses are introduced between the absorber and outside cover, the loss by heat radiation is reduced to approximately

$$q = \frac{1}{\frac{1-\epsilon_1}{\epsilon_1} + (2N-1)\frac{1-\epsilon_2}{\epsilon_2} + N}[\sigma T_1^4 - \sigma T_2^4] \qquad (2\text{–}11)$$

The word "approximately" is used, because, for one thing, allowance for simultaneous conduction or convection through the air between the covers has not been made. Such allowance will be made in the next section. Another factor neglected has been the solar radiation absorbed by the covers. To proceed with our example, if one additional cover is introduced, N = 2, and T_1 and T_2 are imagined as before, the heat flux q drops to

$$q = \frac{1}{\frac{1-0.9}{0.9} + (3)\frac{1-0.8}{0.8} + 2}[463.1] = 162 \ \text{w/m}^2$$

The heat loss is seen to be reduced to less than one half of its former value. Of course, as shown in Table 2-1, the solar energy reaching the absorber is also reduced from 585 w/m^2 to (700)(0.777)(0.95) = 517 w/m^2. A net reduction of (340−162)−(585−517) = 110 w/m^2 in the heat loss is obtained at the economic expense of having to install an additional cover glass.

A minor factor not yet considered is that as additional covers are added the temperature of the outer cover T_2 may drop slightly so that the advantage may not be quite so great as the above example would make it seem, but the solar energy absorbed by the covers, also neglected so far, offsets this minor effect. The point which has been made is that multiple covers offer one means of reducing reradiation heat losses. Of course, each additional cover has a diminishing effect. The reader is invited to use his or her electronic calculator to carry out calculations of q lost by reradiation according to Eq. (2-11) and that reaching the absorber plate using values in Table 2-1. Even when economic considerations are put aside, a point is soon reached when additional covers provide no technical benefit.

Spectral Selectivity. A more effective way to reduce reradiation heat losses is to use a selective black absorber. One contrives to have a low value of ϵ_1 in Eq. (2-10) or (2-11) while at the same time having a high value of solar absorptance α. Any surface whose reflection, transmission, or absorption properties vary with wavelength or photon energy is said to be spectrally selective. A surface which appears black to the eye but has a high reflectance in the infrared spectrum (which would make it appear silvery or metallic to an infrared eye) is said to be a selective black. Here the word "surface" denotes not just an interface but also the layer or layers beneath it.

Let the incoming radiant flux *per unit spectral bandwidth* be $q^{-}(\lambda)$. Spectral location and bandwidth can be specified equivalently by referring to photon wavelength λ, wavenumber ν, frequency ν_f or energy E. They are interrelated by Einstein's photoelectric law $E = h\nu_f$, $\nu = \nu_f/c$, and $\lambda = 1/\nu$. Thus the power per unit area (w/m^2) incident upon a surface and having wavelengths between λ and $\lambda+d\lambda$ is $q^{-}(\lambda)d\lambda$. The quantity $q^{-}(\lambda)$ with units of $\text{w/m}^2\mu\text{m}$ is termed the spectral irradiation. The total irradiation (w/m^2) is the sum over all $d\lambda$ bands of the spectrum.

$$q^{-} = \int_0^\infty q^{-}(\lambda)d\lambda$$

In each spectral region a fraction $\alpha(\lambda)$ of the irradiation is absorbed by a surface. The total flux absorbed is then

$$q_{abs} = \int_0^\infty \alpha(\lambda)q^{-}(\lambda)d\lambda$$

The unitless q_{abs}/q^{-} is the total absorptance of the surface

$$\alpha = \frac{\int_0^\infty \alpha(\lambda)q^{-}(\lambda)d\lambda}{\int_0^\infty q^{-}(\lambda)d\lambda} \tag{2-12}$$

A fractional function may be defined as the fraction of the total irradiation between wavelengths of zero and λ,

$$f(\lambda) = \frac{\int_0^\lambda q^{-}(\lambda)d\lambda}{\int_0^\infty q^{-}(\lambda)d\lambda} \tag{2-13}$$

When this quantity is considered, the total absorptance given by Eq. (2–12) is simply the spectral absorptance $\alpha(\lambda)$ averaged over f.

$$\alpha = \int_0^1 \alpha(\lambda)df(\lambda) \tag{2-14}$$

Figure 2–3 shows a plot of the fractional function $f_S(\lambda)$ for solar energy filtered by a model of the earth's atmosphere. To be noted is the fact that it rises from 2 percent to 98 percent between wavelengths of 0.36 and 2.33 μm, that is, 96 percent of the incident solar energy lies between these two wavelengths.

A surface whose thermodynamic microstate can be characterized by stating that it resembles that existing at thermodynamic equilibrium at temperature T emits spectral flux

$$q_{em}(\lambda) = \alpha(\lambda)B(\lambda,T)$$

where $B(\lambda,T)$ is the Planck black body radiosity

$$B(\lambda,T) = \frac{2\pi hc^2\lambda^{-5}}{e^{hc/\lambda kT}-1} \tag{2-15}$$

The total emittance is then

$$\epsilon(T) = \frac{\int_0^\infty \alpha(\lambda)B(\lambda,T)d\lambda}{\int_0^\infty B(\lambda,T)d\lambda} \tag{2–16}$$

A fractional function for thermodynamic equilibrium heat radiation may be defined as

$$f_e(\lambda T) = \frac{\int_0^\lambda B(\lambda,T)d\lambda}{\int_0^\infty B(\lambda,T)d\lambda} \tag{2–17}$$

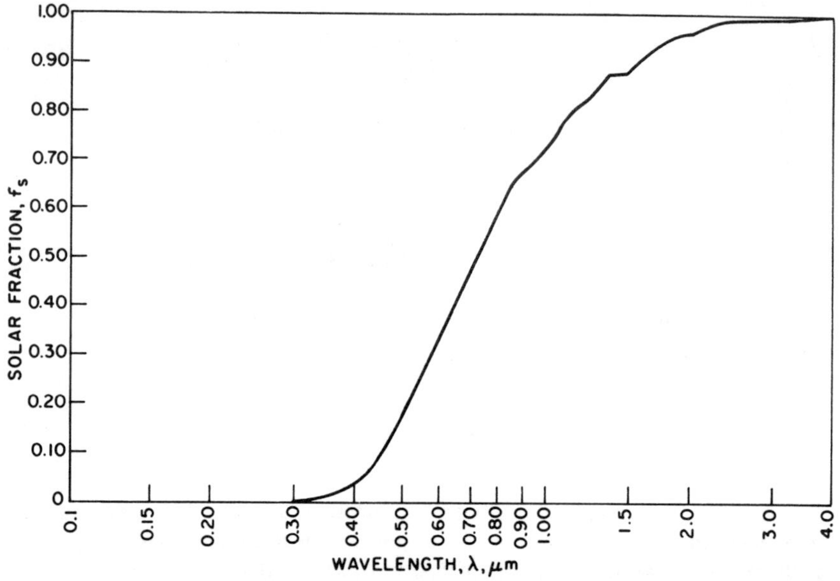

2-3. Fractional Function for Solar Radiation

To distinguish this fraction from a modified one to be defined later, it is called the external fraction. In terms of this function the total emittance $\epsilon(T)$ in Eq. (2–16) becomes simply

$$\epsilon(T) = \int_0^1 \alpha(\lambda)df_e(\lambda T) \tag{2–18}$$

A plot of the external fraction $f_e(\lambda T)$ is shown in Figure 2–4. For a value of

T = 400° K (126.85° C or 260.33° F), it may be noted that 96 percent of the radiation lies between 4 and 44 microns. It is thus possible for a surface to have a high value of $\alpha(\lambda)$ over nearly all of the solar spectrum and simultaneously have a low value in the infrared from 4 to 44 microns.

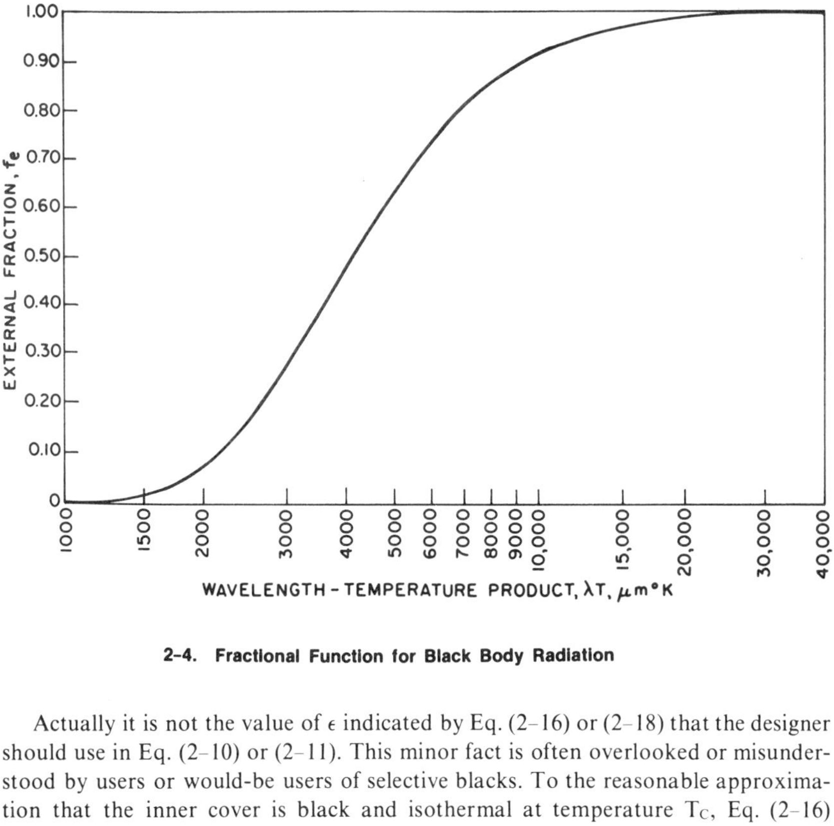

2-4. Fractional Function for Black Body Radiation

Actually it is not the value of ϵ indicated by Eq. (2–16) or (2–18) that the designer should use in Eq. (2–10) or (2–11). This minor fact is often overlooked or misunderstood by users or would-be users of selective blacks. To the reasonable approximation that the inner cover is black and isothermal at temperature T_C, Eq. (2–16) should be written

$$\epsilon(T,T_C) = \frac{\int_0^\infty \alpha(\lambda)[B(\lambda,T)-B(\lambda,T_C)]d\lambda}{\int_0^\infty [B(\lambda,T)-B(\lambda,T_C)]d\lambda} \tag{2–19}$$

and a modified ("internal") fractional function is defined

$$f_i(\lambda T,T_C/T) = \frac{\int_0^\lambda [B(\lambda,T)-B(\lambda,T_C)]d\lambda}{\int_0^\infty [B(\lambda,T)-B(\lambda,T_C)]d\lambda} \tag{2–20}$$

$$\epsilon(T,T_C) = \int_0^1 \alpha(\lambda) df_i(\lambda T, T_C/T) \qquad (2\text{–}21)$$

Figure 2–5 shows a plot of $f_i(\lambda T)$ for $T_C/T \doteq 1$. In this case for $T = 400°K$, the wavelengths encompassing 96 percent of the spectrum of interest lie between 3.4 and 30 microns. Qualitatively speaking, we can state the same conclusion, however. It is possible to have simultaneously a large value of total solar absorptance α_S and a low value of total infrared emittance $\epsilon(T,T_C)$, provided the absorber temperature T is well below the sun's temperature, approximately 5750° K.

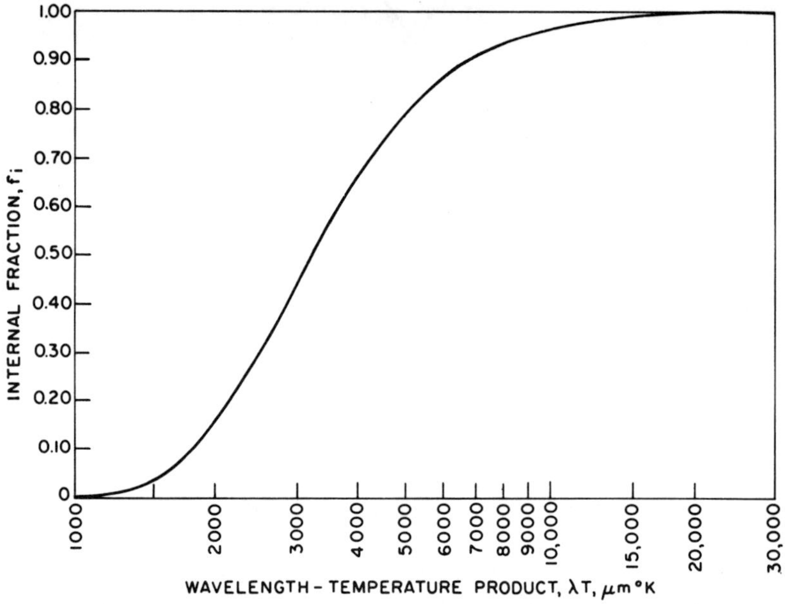

2-5. Fractional Function for a Difference in Black Body Radiation

The wavelength λ_x at which the spectral absorptance of the absorber $\alpha(\lambda)$ should change from a high value to a low value is that at which the solar spectral irradiation $q_S^-(\lambda)$ upon the absorber crosses under the spectral black body radiosity minus the spectral irradiation from the inner coverglass, $B(\lambda,T) - q_C^-(\lambda)$. Ideally the cover glass should transmit only at shorter wavelengths. In other words, one desires

$$\begin{aligned} \alpha(\lambda) &\simeq 1 \qquad \lambda < \lambda_x \\ \alpha(\lambda) &\simeq 0 \qquad \lambda > \lambda_x \end{aligned} \qquad (2\text{–}22)$$

where λ_x is the root of

$$0 = B(\lambda_x,T)-\bar{q_C}(\lambda_x)-\bar{q_S}(\lambda_x) \tag{2–23a}$$

or approximately

$$0 = B(\lambda_x,T)-B(\lambda_x,T_C)-\bar{q_S}(\lambda_x) \tag{2–23b}$$

It should be noted that the solar spectral irradiation $\bar{q_S}(\lambda)$ is that on the absorber after transmission by the cover glasses, that is, after being reduced in the short wavelengths by transmission τ shown in Table 2–1.

Absorptance of Merit. While Eqs. (2–22) and (2–23) state the requirements for an ideal selective black absorber, they are not particularly helpful to the solar designer. In actual practice one has to sacrifice to some extent high values of $\alpha(\lambda)$ at values of λ less than λ_x to obtain low values of $\alpha(\lambda)$ at wavelengths beyond λ_x or vice versa. Thus when one employs a selective black one usually sacrifices a certain $\Delta\alpha_S$ in solar absorptance to obtain a certain $\Delta\epsilon$ decrease in infrared emittance. When is the sacrifice warranted or not? An absorptance of merit incorporating solar absorptance, infrared emittance, and a ratio of solar irradiation to infrared radiation provides the answer.

Again making the approximation that the inner cover is nearly black in the infrared, Eq. (2–10) gives for the heat loss merely

$$q \doteq \epsilon(\sigma T^4-\sigma T_C^4)$$

while the solar heat gained is

$$q_{abs} = \alpha_S \bar{q_S}$$

The net radiation heat flux gained is thus

$$q = \alpha_S \bar{q_S}-\epsilon(\sigma T^4-\sigma T_C^4)$$

In dimensionless terms the fraction of $\bar{q_S}$ retained is $q/\bar{q_S}$. This fraction is called the absorptance of merit,

$$\alpha_M = \alpha_S-\beta\epsilon \tag{2–24}$$

where

$$\beta = (\sigma T^4-\sigma T_C^4)/\bar{q_S} \tag{2–25}$$

The higher α_M the better, technically speaking. Of course, economic considerations dictate finally the choice of absorber surface.

When is a selective surface with, for example, $\alpha_{S,1} = 0.70$, $\epsilon_1 = 0.30$ superior to a black paint with say $\alpha_{S,2} = 0.95$, $\epsilon_2 = 0.88$? Equation (2–24) tells us that the selective surface is superior whenever β exceeds a certain ratio

$$\beta \geqslant (\alpha_{S,2}-\alpha_{S,1})/(\epsilon_2-\epsilon_1)$$

For the example chosen, the value is

$$\beta > (0.95-0.70)/(0.88-0.30) = 0.431$$

To continue the example, how hot must the absorber temperature T be to make the selective black technically desirable? Suppose the cover glass temperature $T_C = 20° C = 293.15° K$ and the solar irradiation q_S^- on the absorber is (700)(0.879) = 615 w/m^2. From Eq. (2–25)

$$\sigma T^4 = \sigma T_C^4 + \beta q_S^-$$

$$\sigma T^4 \geqslant 418.7+(0.431)(615) = 684 \text{ w/m}^2$$

$$T = 331.4° K = 58.3° C$$

In general, spectral selectivity is useful whenever it makes the absorptance of merit α_M greater than that of a competing surface. Such will be the case when the value of β is high, that is, when the collection temperature is relatively high and the solar irradiation relatively low. For flat plate collectors with no concentration a spectrally selective surface is desirable when the absorber temperature T must be say 10 to 40° C above the inner cover glass temperature T_C, the value depending upon the quality of the selective black.

Transmittance of Merit. One may put a spectrally selective coating on one or more of the cover glasses, but when a selective black absorber is used the employment of such a cover glass coating is usually not warranted. However, for windows of habitations or greenhouses such application may be very advantageous, particularly since one obtains the benefit of reducing heat losses during night hours as well as during daylight.

For simplicity consider an absorber with high solar absorptance $\alpha_S \doteq 1$ and high infrared emittance $\epsilon_1 \doteq 1$ underlying a cover with solar transmittance τ_S and infrared emittance ϵ. Such would be the case, for example, for a window opening into a large room; the room would constitute a cavity absorber. Let the solar irradiation on the cover glass be q_S^-. Then the amount of solar energy per unit area (of cover glass) absorbed by the absorber is $\tau_S q_S^-$. The amount lost from the absorber to the cover is $\epsilon(\sigma T^4-\sigma T_C^4)$, so that the net solar energy collected is

$$q = \tau_S q_S^- - \epsilon(\sigma T^4-\sigma T_C^4)$$

A transmittance of merit τ_M may be defined as q/q_S^-

$$\tau_M = \tau_S - \beta\epsilon \tag{2-26}$$

where β is the same quantity defined by Eq. (2–25).

A thin layer of gold (on the order of 100 Angstroms in thickness) with solar antireflection film gives a reasonably high value of τ_S and a low value of ϵ, but the transmittance of merit τ_M tends to be lower than α_M for a selective absorber, because τ_S is lowered somewhat too much. Tin oxide or other semiconductor films also serve, but tend to have the same characteristic, a lower τ_M than the α_M of a good selective black. Thus a selective black absorber appears to be more desirable, but as already mentioned, architectural and agricultural systems may benefit from a selective cover glass.

If one thinks of using a selective cover in conjunction with a selective black absorber, it turns out the added benefit from the selective cover is rather slight while the penalty in loss of transmission is rather high. Thus the combination is not usually attractive.

Selective Blacks. Surfaces with low emittance ϵ and high solar absorptance α_S may be created in a variety of ways. The reader may have experienced the unpleasant sensation of touching unpainted but weathered galvanized iron or aluminum left exposed to the sun. Such a material is a natural selective black and, although of rather poor quality as a selective black, becomes much hotter in the sun than even a black painted surface. Aircraft and bus operators recognize this fact and often paint white the tops of vehicles that may be left parked in the sun before being boarded by passengers. A selective black that is widely available for only a few cents per square meter is thin (say thinner than 25 μm or 0.001 inch thick) black polyethylene, but this material must be backed by a bare metal. It cannot be backed by a high emissivity material or space to be effective. As the absorber for an air heater with the air duct formed between the polyethylene and aluminum foil or aluminized plastic or paper, this material is an excellent selective black but has a limited lifetime. It is susceptible to damage from overheating, too, if the air flow is not adequate.

These examples, weathered galvanized iron and black polyethylene over aluminum foil, exemplify the chief ways selective blacks can be created. In both cases an electrically conducting metal or semiconductor is required. The free electrons in the material create the requisite high infrared reflectance and hence low infrared emittance. In both cases a film transparent to the infrared radiation absorbs the solar radiation. A semiconductor serves admirably, either in the form of a continuous surface film or dispersed in fine particles, a few tenths of angstroms in size. Even a metal may be used in the latter case, for example, gold black; and dielectric films may be used in the former case, as a solar antireflection film on the metal. Polyethylene is rather unique in that it absorbs little infrared radiation outside of two main absorption bands and so may be used to hold dispersed absorbing particles such as the carbon black or soot used to make black film, but polyethylene photodegrades in sunlight. Dispersed particles in a thin single coat of black lacquer on unprimed metal also make a reasonably good selective black. Dispersed particles can be smoked or vapor deposited onto a metal surface without being held in a binder, but adherence and thermal contact (particularly in a vacuum) will be poor.

Just as an antireflection film can be formed on a cover glass by microetching to form a gradient in the effective index of refraction, one can be created on a metal by special microroughness. The roughness elements must be small and closely spaced and rather deep and steep-sided. The gradient in refractive and absorptive index must not be too steep in terms of solar wavelengths, but the entire layer must be thin in terms of infrared wavelengths. Photoetching or deposition can be used to create the requisite microroughness. Again a semiconductor material forming the microroughness on a metal is desirable, but an all-metal surface system can be used.

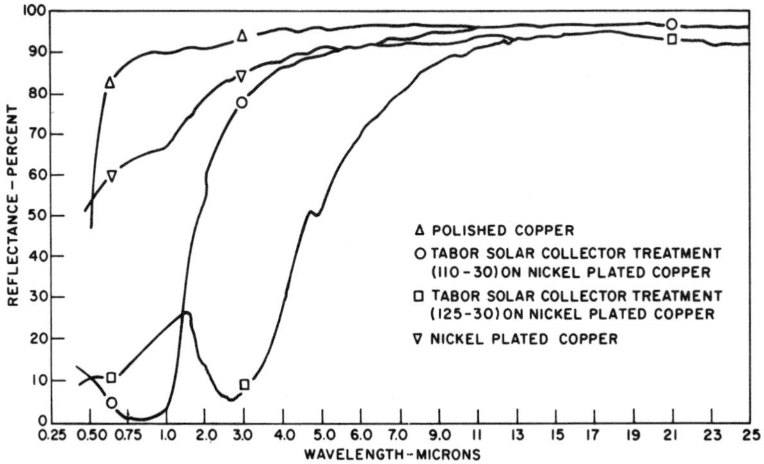

2-6. Spectral Reflectance of Tabor Selective Blacks on Copper

A commercial process to create selective black absorber surfaces is the chemical or electrochemical deposition of a thin semiconductor film on metal. Since 1959 commercial production in Israel has formed such surfaces electrochemically following development by Tabor. Figures 2-6 and 2-7 show the spectral reflectance $\rho(\lambda) = 1-\alpha(\lambda)$ of four Tabor surfaces as well as the untreated surfaces. Note that the absorptance is desirably high (the reflectance low) in the short wavelength solar region and desirably low (the reflectance high) in the long wavelength infrared region. During the film formation, the voltage is varied to obtain an inhomogeneous film with a gradient in the refractive index. The gradient facilitates entry of the solar wavelength photons into the semiconductor where they are absorbed in forming hole-electron pairs. The photon energy is recovered as heat upon recombination of the hole and electron. Absorption of infrared photons is not great, because they do not possess sufficient energy to elevate electrons into the conduction band. Even so, the semiconductor is best kept thin compared to the infrared wavelengths so as to minimize absorption in that region of the spectrum.

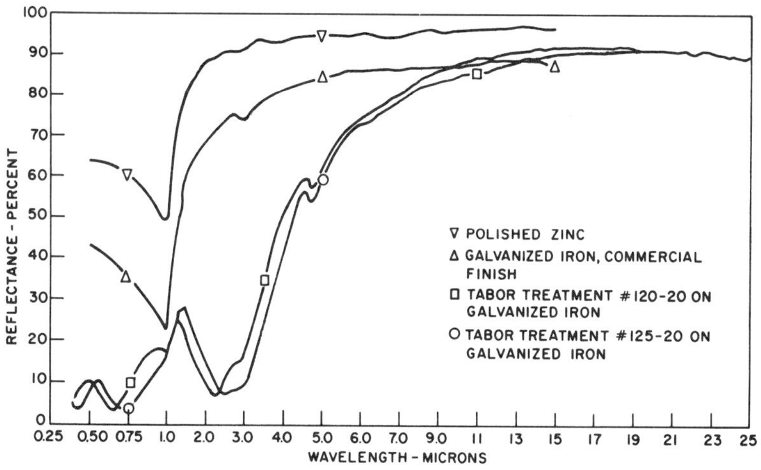

2-7. Spectral Reflectance of Tabor Selective Blacks on Zinc

Chemical dip oxidation treatments can also be employed. Figures 2-8 and 2-9 show such treatments for copper and steel respectively. Oxidation of copper in chlorate solution as a means of obtaining a solar black was proposed by J. T. Gier and R. V. Dunkle in 1955 and brought into commercial practice in Australia as a result of Dunkle's work at the Commonwealth Scientific and Industrial Research Organization.

Directional Selectivity. Because, in a collector with no concentrator, the solar radiation arrives within a cone of small half angle, it is possible to contrive a system which absorbs well any photon coming from the sun's direction but absorbs poorly photons from other directions. Since a surface which absorbs poorly emits poorly, a low emittance for the reradiation is obtained except for the solar directions. The hemispherical emittance, which is a suitably weighted average over all directions, is a good measure of how well or how poorly a surface reradiates. It is defined as

$$\epsilon_H = \frac{1}{\pi}\int_0^{2\pi}\int_0^{\pi/2}\epsilon(\theta,\phi)\cos\theta\sin\theta d\theta d\phi$$

where θ is the angle from the normal to the plane of the mean surface, and ϕ is the azimuthal angle measured in the mean surface plane. It is clear that the directional emittance and absorptance $\epsilon(\theta,\phi)$ may be high for a selected range of directions but the hemispherical value ϵ_H may nevertheless be low.

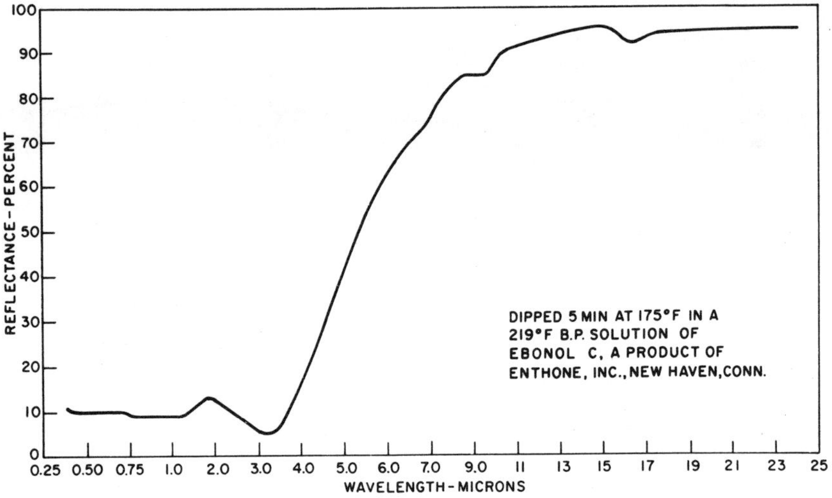

2-8. Spectral Reflectance of Chemically-Oxidized Copper

A vee-groove adds a measure of directional selectivity to a selective black, for example galvanized iron, whose spectral reflectance is shown in Figure 2–7. Radiation coming in from the sun is partially absorbed when it strikes one wall of the groove, and the radiant energy not absorbed is reflected to the other wall which absorbs another portion. If three reflections from the absorbing surface must occur before a loss is suffered as pictured in Figure 2–10, the effective absorptance is $1-\rho^3$ rather than $1-\rho$ when loss occurs after only one reflection. But for angles sufficiently far from the sun only one reflection occurs as shown in the figure. The absorptance and hence emittance for these directions is consequently low. In a fixed collector, the grooves run east-west, and the unit is tilted (south in the Northern Hemisphere) so that the sun's rays are incident from directions of high absorptance. Infrared heat losses are reduced because the infrared heat transfer in directions of low absorptance is reduced.

Thermally-nonconducting and infrared-opaque louvers or honeycomb can likewise provide directional selectivity. In this case rays from the sun may pass through the louvers or honeycomb without reflection or transmission, if the collector tracks the sun, or with a small number of reflections or transmissions if it does not. In the latter case the honeycomb or louvers must be solar reflecting or transmitting. Aluminized paper, plastic overcoated with clear resin for infrared opacity, or thin glass serves well. The solar photons pass through the honeycomb system and reach a high-temperature absorber. Rays away from the sun encounter the outer edges of the honeycomb which are cool, but only when the honeycomb is a poor thermal conductor. The reradiation in these off-solar directions is thus low, and directional selectivity is achieved as shown schematically in Figure 2–11. Honeycomb can simultaneously reduce convection heat losses.

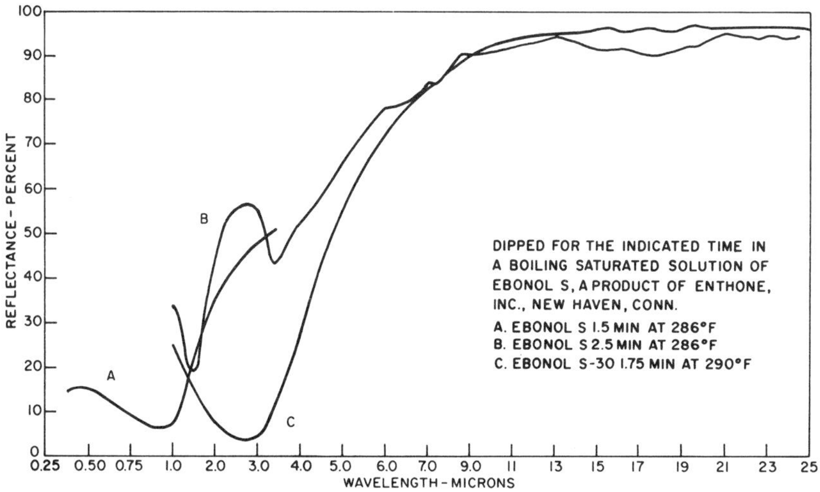

2-9. Spectral Reflectance of Chemically-Oxidized Steel

CONTROL OF CONVECTION

Convection Under a Cover Glass. The conduction heat transfer through the air between a cover glass and absorber (or two cover glasses) a distance L apart is

$$q = (K/L)(T_1 - T_2) \tag{2–27}$$

where K is the thermal conductivity. Recall the example used earlier, where $T_1 = 80°C$ and $T_2 = 20°C$. For $L = 1$ cm $= 0.01$ m the conduction is

$$q = \frac{0.0283}{0.01}(80-20) = 169.7 \text{ w/m}^2$$

Of the 585 w/m^2 absorbed by the absorber in the previous example, the loss of

169.7 w/m^2 would constitute 30%. Recall, however, that the nonselective black previously assumed had a loss of 340 w/m^2 by radiation, so that the net radiant heat collected was only 585−340 = 245 w/m^2. Of this lesser amount the conduction loss would take away nearly 70%. It is thus clear that conduction through the air must be reduced for a flat plate collector with no concentrator.

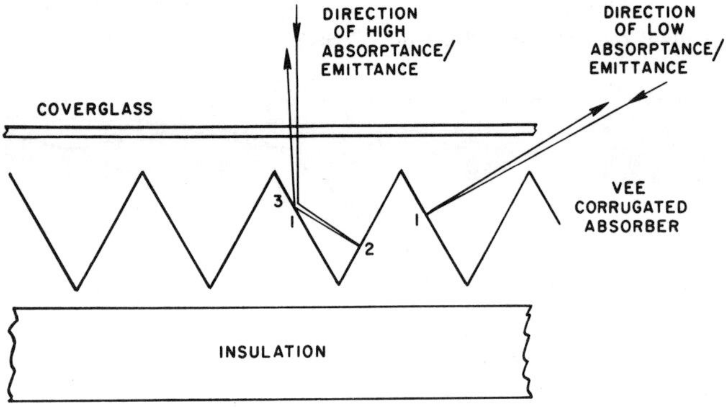

2-10. Ray Traces for a Vee-Groove Absorber

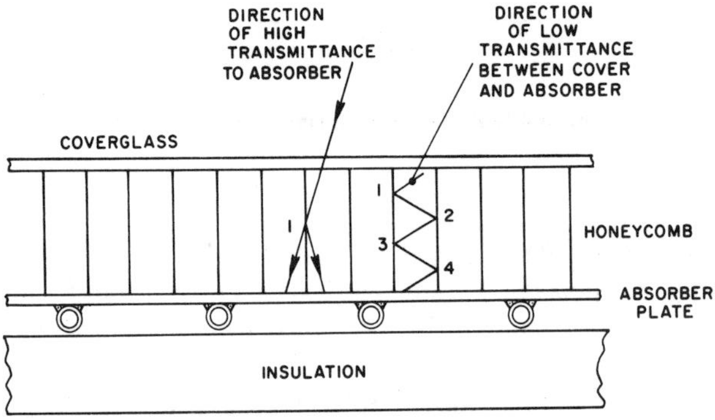

2-11. Ray Traces for a Honeycomb Absorber

Inspection of Eq. (2–27) suggests two possible stratagems, reducing K or increasing L. The former requires replacing the air with another gas, for example, dry carbon dioxide, or using a high vacuum. The latter stratagem seems simple by comparison. But experiment and more sophisticated theory show that increasing L does not bring about the desired reduction that the simple form of Eq. (2–27) suggests. The reason is that warm light air next to the underlying hot absorber surface rises, and the cold heavy air just under the colder cover glass surface falls to replace it. The cold air falling to the vicinity of the hot surface steepens the temperature gradient above the $(T_1-T_2)/L$ value appearing in Eq. (2–27).

The occurrence of natural convection, the movement of air induced by the natural buoyant forces, requires that Eq. (2–27) be modified by multiplying factor greater than or equal to one

$$q = Nu[(K/L)(T_1-T_2)] \tag{2–28}$$

The factor is Nusselt number, a function of Grashof number Gr and Prandtl number Pr. The product of Gr and Pr is Rayleigh number Ra.

$$Gr = \frac{g(\Delta\rho/\rho)L^3}{\nu^2} \tag{2–29}$$

$$Pr = \frac{\nu}{\alpha} = \frac{\mu c_p}{K} \tag{2–30}$$

$$Ra = GrPr = \frac{g(\Delta\rho/\rho)L^3}{\alpha\nu} \tag{2–31}$$

where g is gravitational acceleration, $(\Delta\rho/\rho)$ is the density difference to average density ratio, ν is kinematic viscosity μ/ρ, ρ is density, α is thermal diffusivity $K/\rho c_p$, c_p is specific heat at constant pressure, and K is thermal conductivity. For an ideal gas

$$\Delta\rho \doteq (T_1-T_2)/T \tag{2–32}$$

$$T = (T_1+T_2)/2 \qquad \text{(absolute temperatures)} \tag{2–33}$$

The Nusselt number is also a function of L/D where tilted distance D runs along the absorber plate, angle of tilt τ, and in some cases W/D where W is the width of the absorber plate perpendicular to D.

When D/L and W/L are large and $\tau = 0$ with the hot surface below, theory and experiment show that $Nu = 1$ until Ra exceeds approximately 1700. Thereafter Nu rises, rather steeply at first, but then as $Ra^{1/4}$. The latter variation substituted into Eq. (2–28) shows that q falls with $L^{-1/4}$ as L increases. Thus some benefit is achieved by increasing L but not as much as would be the case had Eq. (2–27) been applicable. Ultimately as L and Ra increase, Nu increases more nearly like $Ra^{1/3}$, and no further benefit is obtained. For this reason a value of L on the order of 8 cm has been suggested. Figure 2–12 shows a plot of conductance $h = q/(T_1-T_2)$ versus distance L for some assumed conditions.

Convection in a Honeycomb. As noted in a previous section, honeycomb can provide directional selectivity and thus reduce radiation heat losses. Consideration of the radiation transfer as sketched in Figure 2–11 indicates that only the ratio L/D is important. For example, if a value of L/D = 10 is desired, the radiation transfer equations are indifferent as to whether D is 1 cm or 10 cm. A value of D on the order of 4 cm might seem desirable from a point of view of fabrication. However, if one considers the natural convection and wishes to suppress it, the value of D becomes very critical. As D is reduced the natural convection flow current has greater mechanical energy dissipated by viscosity, the thermal energy is drained from it by increased lateral conduction. Thus the stability of the fluid in the honeycomb is increased.

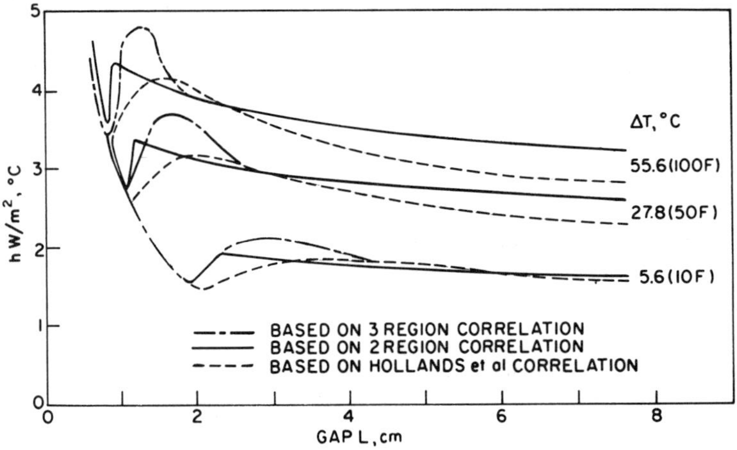

2-12. Conductive-Convective Heat Transfer Across a Gap

Mathematically, the critical Rayleigh number based upon D that describes the conditions under which natural convection initiates varies only weakly with L/D and W/D. Hence the temperature difference which can be borne across the absorber-cover glass gap rises greatly with decreasing D

$$Ra_D = Ra(L/D)^4 = \frac{g(\Delta T/LT)D^4}{\alpha\nu}$$

$$\Delta T = \frac{\alpha\nu LT}{gD^4} Ra_D$$

If one wants to achieve a value of $\Delta T = 100°C$, one must have a value of D on the order of 0.5 cm for a rectangular honeycomb or 1 cm for a hexagonal one. Figure 2-13 shows a plot of D versus ΔT for a rectangular honeycomb. For a hexagonal honeycomb D can be approximately twice as large.

When Ra exceeds the critical value the Nu factor increases very rapidly, so that the honeycomb rapidly loses its ability to suppress natural convection.

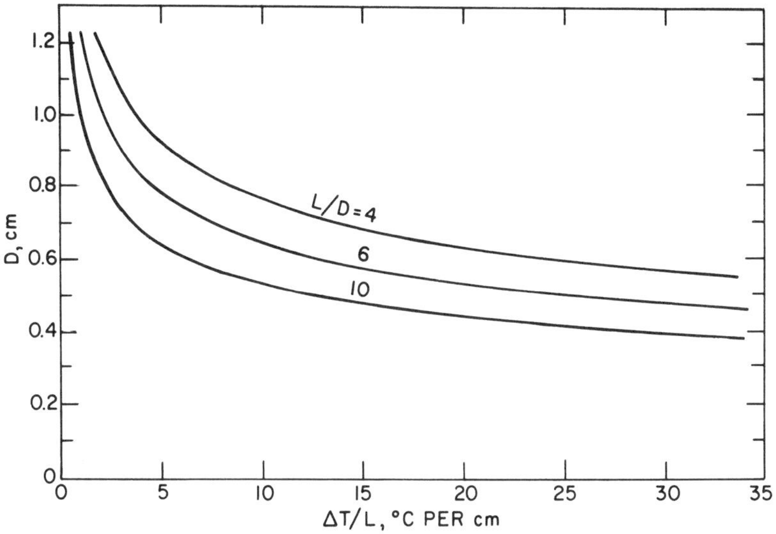

2-13. Dimension D versus ΔT/L for a Rectangular Honeycomb

THERMAL INSULATION

The foregoing examples have shown that a loss of 100 or 200 w/m^2 can have very serious consequences for a flat plate solar collector with no concentrator. For this reason it is usually desirable to insulate the back or bottom of the collector with a material of low thermal conductivity K and low specific heat and density. To achieve a loss of one-tenth of the above values, a thickness L_i of insulation is indicated from Eq. (2-27) which applies. For a fibrous or foam insulation with $K = 0.06\ w/m\,^\circ K$, a thickness is needed of

$$L_i = (K/q)(T_1 - T_2) = (0.06/20)(60) = 0.18 \text{ m}$$

This thickness amounts to 18 cm or approximately 7 inches. It is important that the insulation be kept dry. Of course, if the value of $(T_1 - T_2)$ desired is smaller and/or a greater q lost is acceptable, the thickness L_i can be reduced accordingly. Thicker amounts of insulation are not beneficial, not only because the residual reductions in heat loss are small, but also because the daily transient nature of the solar energy is such that the outer layers of the insulation do not provide benefit during warm up.

One or more air gaps lined with metal foil beneath the absorber form an inexpensive, low-mass heat barrier. Such a barrier may be used even when fibrous or foam insulation is used below it, because it protects the insulation from the highest temperatures next to the absorber.

Penetrations through the insulation for structural support or piping connections are a matter of serious concern. Needless to say, they should be made of poor conductors and as small in area as permissible. Wood, for example, is a fairly poor conductor and desirable structural material; it conducts only two or three times as well as fiberglass insulation. Concrete conducts 20 times as well, and steel 1000 times. Penetrations by these latter materials are to be made as small in cross-sectional area as possible.

3. HEAT COLLECTION

THE SOLAR COLLECTOR AS A HEAT EXCHANGER

Heat Collection by an Element. Consider an element of solar collector as sketched in Figure 3–1. The working fluid, usually air or water, enters and leaves the element at the rate of $\dot{m}$, kg/sec or lb/hr. If the fluid remains in a single gas or liquid phase (that is, the water does not boil) the amount of heat power $d\dot{Q}$, w or Btu/hr, collected by the fluid is related to the temperature rise of the fluid dT by

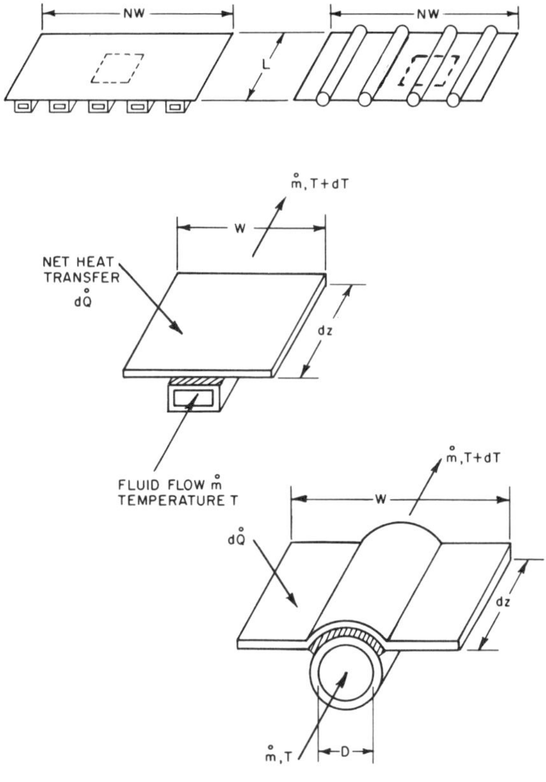

3-1. An Element of a Flat-Plate Solar Collector

$$d\dot{Q} = \dot{m}c_p dT \tag{3–1}$$

where c_p is the specific heat at constant pressure in j/kg° C or Btu/lb° F.

Air: $c_p \doteq 1000$ j/kg° C, 0.24 Btu/lb° F

Water: $c_p \doteq 4200$ j/kg° C, 1 Btu/lb° F

For example, a power $d\dot{Q} = 1000$ watts will raise the temperature of air flowing at $\dot{m} = 1$ kilogram per second by 1 Celsius degree (1 watt-second equals 1 joule) or 3412 Btu/hr will heat 7937 lb/hr of air by 1.8 Fahrenheit degrees. The heat collected is the solar power absorbed by the absorber less that lost,

$$d\dot{Q} = \alpha_S \tau_S (1+a) \bar{q_S} dA - U dA (T - T_e) \qquad (3\text{–}2)$$

where α_S is the solar absorptivity of the absorber, τ_S the solar transmissivity of the cover, *a* the cover glass absorption recoup, $\bar{q_S}$ the solar irradiation (w/m^2), dA is the elemental area, U the overall heat loss conductance, and T_e the equilibrium temperature of the collector when shaded from the sun (but not the sky). The area dA is the width of the element w times its elemental length dz

$$dA = W dz \qquad (3\text{–}3)$$

As a matter of some convenience there is defined an equivalent static temperature T_s

$$\alpha_S \tau_S (1+a) \bar{q_S} + U T_e = U T_s \qquad (3\text{–}4)$$

For example, if α_S is 0.95, τ_S is 0.88, a is 0.024, $\bar{q_S}$ is 665 w/m^2 (211 Btu/hr ft^2), U is 6 $w/m^{2\circ}C$ (1.06 Btu/hr $ft^{2\circ}F$), and T_e is 15° C; then T_s is 109.8 (230° F). With this convenient definition, Eq. (3–1) becomes

$$\dot{m} c_p \frac{dT}{dz} = U W (T_s - T) \qquad (3\text{–}5)$$

Effectiveness and Number of Transfer Units. Equation (3–5) may be solved for the outlet temperature T_o at $z = L$ when the inlet temperature is specified to be T_i at $z = 0$. The relation resulting may be written in the form

$$T_o = T_i + (T_s - T_i) E_{he} \qquad (3\text{–}6)$$

where E_{he} is the heat exchanger effectiveness, which in turn is given by

$$E_{he} = 1 - e^{-N_{tu}} \qquad (3\text{–}7)$$

where N_{tu} is the so-called number of transfer units

$$N_{tu} = \frac{UWL}{\dot{m} c_p} \qquad (3\text{–}8)$$

To make use of these relations in a design sense, Eqs. (3–6) and (3–7) are better rearranged in the form

$$E_{he} = \frac{T_o - T_i}{T_s - T_i} \tag{3-9}$$

$$N_{tu} = ln \frac{1}{1 - E_{he}} \tag{3-10}$$

For example, suppose one would like to know how much water should flow through a collector per elemental width W = 0.25 m which is L = 1 meter long when it is desired that the temperature rise from T_i = 20° C to T_o = 60° C. We use U = 6 $w/m^{2\circ}C$ and T_s = 109.8° C as in our previous example. One begins by computing the desired effectiveness from Eq. (3–9).

$$E_{he} = \frac{60-20}{109.8-20} = 0.445$$

and then finds the number of transfer units from Eq. (3–10)

$$N_{tu} = ln \frac{1}{1-0.445} = 0.590$$

From Eq. (3–8) the flow rate is determined

$$\dot{m} = \frac{UWL}{N_{tu}c_p} = \frac{(6)(0.25)(1)}{(0.590)(4200)} = 0.606\text{x}10^{-3}\ \text{kg/sec}$$

$$\dot{m} = (0.606\text{x}10^{-3}\ \tfrac{\text{kg}}{\text{sec}})(3600\ \tfrac{\text{sec}}{\text{hr}}) = 2.18\ \tfrac{\text{kg}}{\text{hr}} \qquad (4.81\ \tfrac{\text{lb}}{\text{hr}})$$

If one needs 72 kg/hr (approximately 19 gal/hr) of water, one needs 72/2.18 = 33 W-wide elements.

Calculation of U and T_e. While U and T_e are both rather easily measured, the designer needs to know how they may be influenced by design variables. Before attacking this problem, consider how the quantities could be measured. With the collector shaded from the sun but exposed to the sky and with zero flow rate $\dot{m}$ = 0, the temperature T_e is measured. Then hot water is delivered at T_i, and the colder outlet temperature T_o is observed as well as the flow rate $\dot{m}$. The flow rate may be observed by seeing the time necessary for a container on a weigh-scale to collect a certain mass. With the known inlet and outlet temperatures, and with the knowledge that T_s = T_e because of the shading, Eq. (3–9) is used to find the value of E_{he} appropriate to the test conditions. Equation (3–10) then gives N_{tu}, and Eq. (3–8) is solved for $U = \dot{m}c_pN_{tu}/WL$.

To attack the problem of deriving U and T_e in terms of the environmental conditions and design variables, one writes the heat balances for the absorber and each cover glass. For the absorber the useful heat flux q_u is that absorbed q_a minus that lost q_l at steady state.

$$q_u = q_a - q_l \tag{3-11}$$

We have found that

$$q_l = q_r + q_c + q_k \tag{3-12}$$

where q_r is the radiant heat flux lost to the cover given by Eq. (2–10) for parallel plates, q_c is the convective heat flux given by Eq. (2–28), and q_k is the conductive flux given by Eq. (2–27). With reference to these equations one defines

$$h_r = \frac{1}{\dfrac{1-\epsilon_1}{\epsilon_1} + \dfrac{1-\epsilon_2}{\epsilon_2} + 1}\,\sigma(T_1^2+T_2^2)(T_1+T_2) \tag{3-13}$$

$$h_c = Nu\,\frac{K}{L_c} \tag{3-14}$$

$$U_{ins} = \left(\frac{K}{L}\right)_{ins} \tag{3-15}$$

$$U_i = h_r + h_c \tag{3-16}$$

For example, with $\epsilon_a = 0.88$, $\epsilon_2 = 0.80$, $T_1 = 335°K$, $T_2 = 310°K$, $L_c = 6$ cm, $K_{ins} = 0.06$ w/m°C, and $L_{ins} = 18$ cm, the value of U_i is

$$U_i = 5.5+3 = 8.5 \text{ w/m}^{2\circ}\text{C } (1.50 \text{ Btu/hr ft}^{2\circ}\text{F})$$

and

$$U_{ins} = 0.06/0.18 = 0.333 \text{ w/m}^{2\circ}\text{C } (0.06 \text{ Btu/hr ft}^{2\circ}\text{F})$$

The quantity q_a is, with the small correction a_r for interreflections,

$$q_a = \frac{\alpha_S \tau_S \bar{q_S}}{1-\rho_S(1-\alpha_S)} = \alpha_S \tau_S (1+a_r)\bar{q_S} \tag{3-17}$$

where τ_S is the solar transmittance of the cover system given by Table 2–1, ρ_S is its solar reflectance, and α_S is the absorber solar absorptivity. For example, with $\alpha_S = 0.95$, $\tau_S = 0.879$, and $\rho_S = 0.0735$, the quantity $a_r = 0.0037$. Equation (3–11) becomes

$$\alpha_S \tau_S (1+a_r)\bar{q_S} - q_u = U_i(T_1-T_2) = U_{ins}(T_1-T_e) \tag{3-18}$$

where the rather inconsequential approximation has been made that the heat sink for the insulation is at temperature T_e.

The heat balance on the outer coverglass at temperature T_n (for a single cover n = 2) states that the heat gain from the interior $U_{i,n}(T_{n-1}-T_n)$ and the absorbed solar and sky irradiation $\alpha_g \bar{q_S} + \epsilon_g \bar{q}_{IR}$ must equal the heat lost to the sky by radiation $\epsilon_g \sigma T_n^4$ and convection $h_o(T_n-T_a)$ where T_a is the air temperature

$$U_{i,n}(T_{n-1}-T_n) + \alpha_g \bar{q}_S + \epsilon_g \bar{q}_{IR} = \epsilon_g \sigma T_n^4 + h_o(T_n-T_a) \qquad (3\text{–}19)$$

The quantity α_g is the solar absorptivity of the cover augmented slightly by interreflections. The sky irradiation at sea level is given approximately by the Brunt equation

$$\bar{q}_{IR} \doteq (0.55+1.8\sqrt{P/P_o})\sigma T_a^4 \qquad (3\text{–}20)$$

where P_o is one atmosphere and P is the partial pressure of the water vapor in the atmosphere. In terms of the relative humidity r_H

$$P = r_H P_{vap}(T_a) \qquad (3\text{–}21)$$

where $P_{vap}(T_a)$ is the vapor pressure of steam at temperature T_a. For example, at 20° C and 50 percent relative humidity the infrared irradiation from the sky is

$$\bar{q}_{IR} = (0.55+1.8\sqrt{0.0115})(5.6697\text{x}10^{-8})(293.15)^4 = 311 \text{ w/m}^2 \text{ (98.6 Btu/hr ft}^2\text{)}$$

It is Eq. (3–19) that leads to the definition of T_e. The right hand side of the equation is written

$$\epsilon_g \sigma T_n^4 + h_o(T_n-T_a) = \epsilon_g \sigma T_a^4 + U_o(T_n-T_a)$$

where

$$U_o = h_o + \epsilon_g \sigma (T_n^2+T_a^2)(T_n+T_a) \qquad (3\text{–}22)$$

and

$$U_o T_a + \epsilon_g \bar{q}_{IR} - \epsilon_g \sigma T_a^4 = U_o T_e$$

$$T_e = T_a - \frac{\epsilon_g \sigma T_a^4 - \epsilon_g \bar{q}_{IR}}{U_o} \qquad (3\text{–}23)$$

For example, with h_o (which must be measured or calculated based upon the prevailing wind conditions or based upon handbook values) equal to say 12 w/m^2°C (2.12 Btu/hr ft^2°F), T_n = 310° K, T_a = 293° K, ϵ_g = 0.80, and $\bar{q}_{IR}$ = 311 w/m^2 as in the previous situation, one obtains from Eq. (3–22) U_o = 12+5 = 17 w/m^2°C and from Eq. (3–23) T_e = 20−5 = 15° C.

With these definitions, Eq. (3–19) can be written simply

$$U_{i,n}(T_{n-1}-T_n) + \alpha_g \bar{q_S} = U_o(T_n-T_e) \tag{3–24}$$

For inner cover glasses, if any,

$$U_{i,m}(T_{m-1}-T_m) + \alpha_{g,s} \bar{q_S} = U_{i,m+1}(T_m-T_{m+1}) \tag{3–25}$$

Equations (3–18), (3–24), and (3–25) may be solved simultaneously for T_1 to T_n. For a single cover glass, n = 2, Eq. (3–25) does not pertain, and Eqs. (3–18) and (3–24) are solved simultaneously for T_1 and T_2. With the known temperatures, q_u is found and $d\dot{Q} = q_u dA$ is compared with Eq. (3–2) to find U

$$U = U_{ins} + \frac{1}{\dfrac{1}{U_o} + \dfrac{1}{U_1}} \tag{3–26}$$

and simultaneously

$$a = a_r + \frac{\alpha_g}{\tau_S \alpha_S} \frac{U_i}{U_o+U_i} \tag{3–27}$$

For example, with U_{ins} = 0.333, U_i = 8.5, and U_o = 17 w/m^{2o}C, there results U = 6.0 w/m^2 (1.06 Btu/hr ft^{2o}F) and with α_g = 0.05, τ_S = 0.879, and α_S = 0.95, there is found a = 0.024.

Fin Effectiveness and Thermal Contact. If the entire absorber surface is not wetted on the back of the W-wide element, the element must act as a fin to conduct thermally the heat to the passage in which the working fluid flows. The thermal contact between the absorber plate and flow passage is then a consideration. The value of U appearing in Eq. (3–8) must be reduced by an effectiveness factor η_f accounting for these effects. The following analysis illustrates how η_f is calculated for the geometry shown in Figure 3–2.

Following the procedures developed in any heat transfer text,* one might idealize the situation as two fins of length *l*, perimeter Δz, and cross-sectional area $b\Delta z$ joined to an isothermal plate of area $W_o\Delta z$ conducting heat through a slab of area $W_o\Delta z$ and thickness δ to an isothermal tube having inside area $p_i\Delta z$ where p_i is the wetted perimeter

$$p_i = 2W_i + 2H_i \tag{3–28}$$

On the area $p_i\Delta z$ heat transfer coefficient h_i acts. The coefficient h_i must be calculated according to the procedure shown in a heat transfer text. For water flowing at

*For example, *Transfer Processes,* by D. K. Edwards, V. E. Denny, and A. F. Mills, Hemisphere/McGraw-Hill, New York, 1976.

2.18 kg/hr in a 12x24 mm duct, h_i is approximately 170 w/m$^{2\circ}$C. The thermal circuit for this situation is shown in Figure 3–3. From it one has

$$\frac{1}{\eta_f UW} = \frac{1}{2\sqrt{UK_w b}\,\tanh\sqrt{Ul^2/K_w b} + UW_o} + \frac{\delta}{K_a W_o} + \frac{1}{h_i p_i} \qquad (3\text{–}29)$$

Conventional practice for a solar water heater (recall Figure 1–2) would be to use a spacing between tubes of W = 0.12 m together with a copper absorber plate thick-

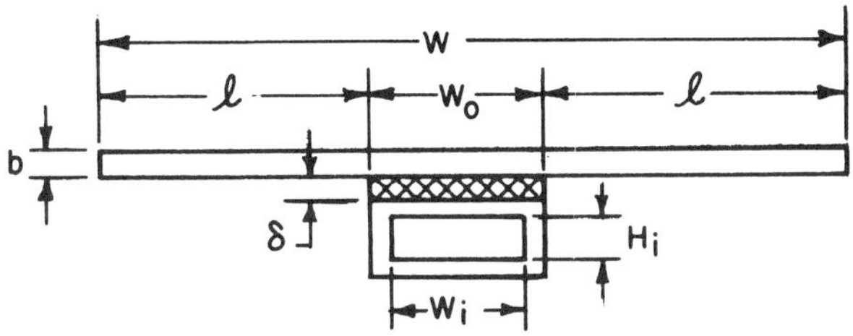

3-2. A Collector Element Viewed as a Fin

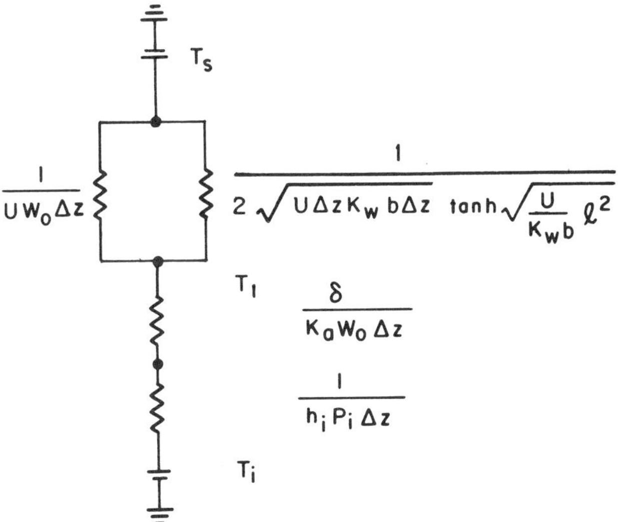

3-3. An Approximate Thermal Circuit for the Collector Fin

ness b of 0.3 mm. In the example taken here, a larger value of W and a thinner value of b are chosen to emphasize their effect. We take the following values:

$$U = 6 \ w/m^{2\circ}C, \qquad W = 0.25 \ m$$

$$b = 0.25 \ mm, \qquad K_w = 380 \ w/m^\circ C \ (\text{copper})$$

$$l = 0.1123 \ m, \qquad W_o = 0.0254 \ m$$

$$\delta = 0.12 \ mm, \qquad K_a = 0.4 \ w/m^\circ C \ (\text{dielectric-filled plastic adhesive})$$

$$h_i = 170 \ w/m^{2\circ}C, \qquad p_i = 0.072 \ m$$

We obtain

$$\frac{1}{\eta_f U W} = 0.814 + 0.0118 + 0.0817 = 0.908$$

$$\eta_f = 0.7346$$

In this example the effective value of U is reduced from 6.0 $w/m^{2\circ}C$ to $\eta_f U$ = 4.41 $w/m^{2\circ}C$, the reduction due in the main to the thermal resistance of the thin copper fin. The width or tube spacing W is thus seen to be too large for the thin b-value chosen for the example. Table 3–1 shows values for other values of W and b.

To appreciate more fully the fin effectiveness role, we can recompute $\dot{m}$ from Eq. (3–8). The loss of flow for specified inlet and outlet temperature and collector size is the loss of capacity. The values obtained are as follows:

TABLE 3–1

FIN EFFECTIVENESS VERSUS COLLECTOR CONDUIT SPACING W AND ABSORBER PLATE THICKNESS b

($U = 6 \ w/m^{2\circ}C$, $K_w = 380 \ w/m^\circ C$, $W_o = 0.0254$ m, $\delta = 0.12$ mm, $K_a = 0.4 \ w/m^\circ C$, $h_i = 170 \ w/m^{2\circ}C$, $p_i = 0.072$ m)

Fin Thickness b (mm)	Fin Effectiveness for the Value of Spacing W (cm)								
	W = 5	7.5	10	12.5	15	17.5	20	22.5	25
0.15	.9703	.9468	.9160	.8794	.8390	.7965	.7535	.7111	.6703
0.20	.9709	.9500	.9234	.8922	.8577	.8210	.7833	.7455	.7084
0.25	.9712	.9519	.9279	.9002	.8695	.8369	.8030	.7687	.7346
0.30	.9715	.9531	.9310	.9056	.8777	.8479	.8169	.7854	.7537
0.35	.9717	.9541	.9332	.9096	.8837	.8561	.8274	.7980	.7683
0.40	.9718	.9547	.9349	.9126	.8883	.8624	.8354	.8078	.7799
0.45	.9719	.9553	.9362	.9149	.8919	.8674	.8419	.8157	.7892
0.50	.9720	.9557	.9372	.9168	.8948	.8714	.8471	.8222	.7968

Case (a) Thick Absorber Plate or Wide Flow Duct	Case (b) Thin Absorber Plate and Narrow Flow Duct
$\eta_f = 1.0$	$\eta_f = 0.735$
$U = 6.0\ w/m^{2\circ}C$	$U = 6.0\ w/m^{2\circ}C$
$\eta_f U = 6.0\ w/m^{2\circ}C$	$\eta_f U = 4.41\ w/m^{2\circ}C$
$\dot{m} = 2.18$ kg/hr (per tube)	$\dot{m} = 1.60$ kg/hr (per tube)

We see that the loss of flow is directly proportional to the effectiveness, so that the quantity is quite important.

It can be noted in passing that for the large cover glass spacing $L_c = 6$ cm chosen here the loss of effectiveness would not be quite so great because of radiation and conduction heat transfer along the cover. Here for simplicity only transfer through the cover was analyzed. However, the qualitative conclusion is correct that too wide spacings W, too narrow channels W_o, too thin a plate thickness b, and too thick an adhesive δ of too poor a conductivity K_a will result in serious loss of effectiveness.

Collector Efficiency. The instantaneous collector efficiency is the instantaneous power collected divided by the solar power received on the collector area.

$$\eta = \frac{\dot{m} c_p (T_o - T_i)}{\bar{q_s} W L} \qquad (3\text{–}30)$$

For example, in the situation described above for case (a)

$$\eta = \frac{(2.18/3600)(4200)(60-20)}{(665)(0.25)(1.00)} = 0.612$$

Sixty-one percent of the solar power incident upon the collector is transferred to the working fluid in this example.

Some algebraic manipulation puts Eq. (3–30) into a form in which the influence of the outlet temperature T_o, the heat loss coefficient U, and the level of solar irradiation $\bar{q_s}$ can be more readily seen. Equation (3–8) is used to eliminate $\dot{m}c_p$ in favor of N_{tu}; Eq. (3–10) is used to eliminate N_{tu} in favor of the inlet, outlet, and stagnation temperatures T_i, T_o, and T_s; and Eq. (3–4) is used to eliminate T_s. Note that the U in Eq. (3–8) must be modified by the fin effectiveness η_f, but not the U in Eq. (3–4). With this manipulation the instantaneous efficiency η becomes

$$\eta = \frac{\eta_f[U(T_o - T_i)/\bar{q_S}]}{ln\left[1 + \frac{[U(T_o - T_i)/\bar{q_S}]}{[\alpha_S \tau_S(1+a)] - [U(T_o - T_e)/\bar{q_S}]}\right]} \tag{3-31}$$

Note that the fin efficiency is a multiplicative factor

Under the special conditions that the outlet temperature is removed from the stagnation temperature and the inlet temperature is not far removed from the outlet temperature so that the ratio $(T_o - T_i)/(T_s - T_o)$ is small, the fact that $ln(1+x) \doteq x$ when x is small shows that

$$\eta = \eta_f \alpha_S \tau_S (1+a) - \eta_f U(T_o - T_e)/\bar{q_S} \tag{3-32}$$

This expression is the basis for plotting instantaneous efficiency versus $(T_o - T_e)/\bar{q_S}$ and expecting a nearly-straight-line relationship. The slight nonlinearity is introduced primarily by the variation of U with temperature. Figure 3–4 shows such plots for three solar collectors. In the plot it is obvious that a honeycomb collector, because of the low value of heat loss coefficient U, is the best suited for high temperature operation.

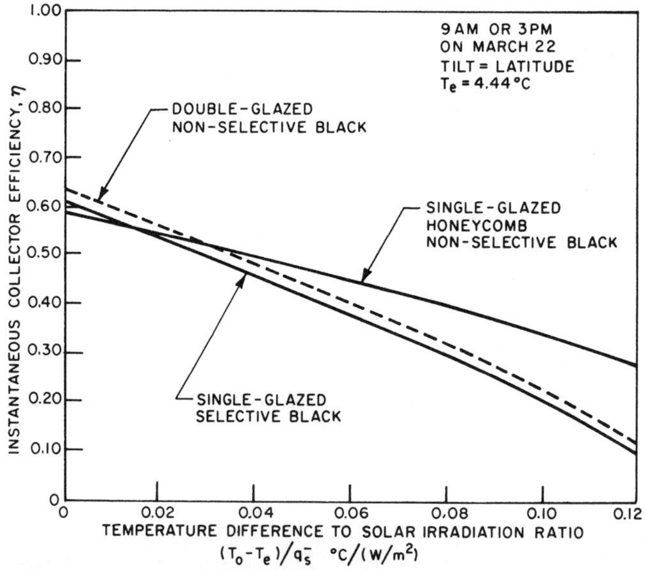

3-4. Instantaneous Efficiency Versus Temperature-Difference-to-Solar-Irradiation Ratio

WORKING FLUID FLOW

Flow Requirements. In the previous section, the mass flow rate desired was shown to be fixed by an inlet and outlet temperature specification. Two main alternatives are used to obtain the desired flow: (1) a circulating pump or (2) a thermosyphon. If the first alternative is employed, it is necessary to have a pump capable of delivering the required flow (19 gal/hr in our previous example) at the pressure drop imposed by the flow resistance of the system. Pump selection is well understood by Mechanical and Chemical Engineers, and will not be reviewed here. A thermosyphon can act when cold working fluid is available at a higher elevation than the solar collector. Such a means of providing the flow requirement is particularly convenient when it can be used, because no motor, pump, or power supply is needed.

Open and Closed Flow Systems. Whether a pump or thermosyphon is used, the working fluid can flow in an open or closed system. In an open system the working fluid is the fluid that is desired to be heated, the hot water itself in the case of a domestic water heater, or the feedwater and steam in the case of a solar-steam-turbine-electric-generator system. In a simple (perhaps intermittent or batch) water-preheater, line water may flow into and out of the solar collector on its way to the main heater and storage tank as hot water is drawn from the latter. In a slightly more sophisticated arrangement, cold water from the bottom of a storage tank may be circulated through the collector to return to the top of the storage tank. Hot water is drawn off the top as required and cold water replaces it at the bottom and is heated by circulating through the solar unit. A thermostat or solar cell is normally used to turn the pump on and off.

In a closed system the working fluid is circulated through a closed flow loop and does not mix with the fluid being heated. For example, a water-glycol mixture (to prevent damage from freezing) may be circulated through the solar collector and then through a coil-heat-exchanger at the bottom of a hot water storage tank. This circulation may be forced by a pump or by a natural thermosyphon effect.

The Thermosyphon. In the thermosyphon the pressure ΔP necessary to force the desired rate of flow $\dot{m}$ is provided by gravity g acting upon the density difference $\rho_C - \rho_H$ between the cold supply fluid in the downcomer pipe and the hot fluid in the riser over the height of the downcomer H, as shown in Figure 3–5.

$$\Delta P(\dot{m}) = (\rho_C - \rho_H)gH \qquad (3\text{–}33)$$

The pressure-flow-rate relation $\Delta P(\dot{m})$ is strongly influenced, of course, by the sizes and lengths of the flow elements, the smoothness of their walls, the gradualness of the transitions from one size to another, and the gentleness of curvature around corners and so forth. The sizes of the passages must be large enough or H high enough so that the desired flow rate $\dot{m}$ is obtained. Needless to say, the passages should be arranged to purge themselves of trapped bubbles or air pockets.

As a concrete, even if oversimplified example, suppose that 2.18 kg/hr (0.58) gal/hr) must flow through a smooth tube with inside diameter D, to be determined, and length L = 1 meter. The tube may be one of a number as pictured in Figure 3–1. The pressure drop-flow relation is given (parametrically in velocity v) by the following relations:

$$\Delta P = (f\frac{L}{D} + K_i + K_o)\frac{1}{2}\rho v^2 \qquad (3\text{–}34)$$

$$f = f(Re) \qquad (= 64/Re,\ Re < 2300) \qquad (3\text{–}35)$$

$$Re = \frac{vD}{\nu} \qquad (3\text{–}36)$$

$$\dot{m} = \rho v(\pi D^2/4) \qquad (3\text{–}37)$$

where f is the friction factor, K_i and K_o are inlet and outlet loss coefficients respectively, ρ is density, Re is Reynolds number, and ν is kinematic viscosity. Suppose $H = 2$ meters and $(\rho_C - \rho_H)$ is that appropriate to hot water at $T_H = 60°C$ and cold water at $T_C = 20°C$. From Eq. (3–30) the pressure available is

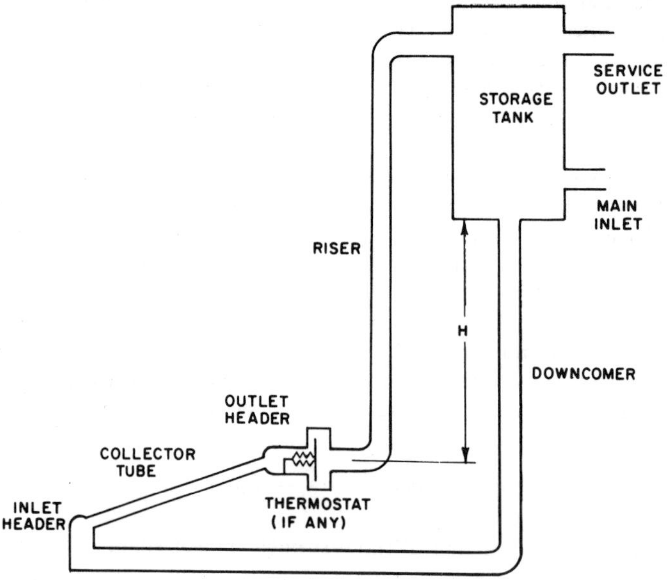

3-5. An Open Thermosyphon

$$\Delta P = (1000.52 - 985.46)(9.807)(2) = 295\ N/m^2$$

or

$$= (62.46 - 61.52)(6.56) = \qquad 6.17\ lb_f/ft^2$$

Equation (3–34) then gives the pressure required, which must not exceed that available. One chooses D, computes v from Eq. (3–37), Re from Eq. (3–36), f from Eq.

(3–35) and finally ΔP from Eq. (3–34). A larger value of D is chosen if ΔP from Eq. (3–34) is too high. This procedure yields an inside diameter D = 2.74 mm (0.108 inch) for the example chosen here (using a mean value of $\nu = 0.658 \times 10^{-6}$ m^2/sec, $\rho = 994.6$ kg/m^3, $K_i = 0.5$, and $K_o = 1$).

In the preceding example only the pressure loss through a single tube was equated to the drive pressure available according to Eq. (3–33). Actually all the pressure drops of the flow around the flow circuit in Figure 3–5 have to be added; that is, the right hand side of Eq. (3–34) in general contains a sum of such terms as the one shown.

For another example, suppose one wanted $T_i = 50°$ C and $T_o = 60°$ C, with the other values previously chosen unchanged. In this case, from Eqs. (3–9), (3–10), and (3–8)

$$E_{he} = \frac{60-50}{109.8-50} = 0.167$$

$$N_{tu} = ln \frac{1}{1-0.167} = 0.183$$

$$\dot{m} = \frac{(6)(0.25)(1)}{(0.183)(4200)} = 1.95 \times 10^{-3} \text{ kg/sec}$$

$$\dot{m} = 7.03 \text{ kg/hr}$$

In this case the flow rate desired is seen to be approximately three times greater than that for the lower inlet temperature, and the ΔP available according to Eq. (3–33) will be approximately one-fourth as much, because $T_o - T_i = 10°$ C rather than 40° C. In order to achieve 3 times as much flow with one-fourth as much pressure, one needs a considerably larger value of inside diameter D, a value on the order of 4.8 mm instead of 2.7 mm for the lower inlet temperature. The heaters, riser, and downcomer which connect to the many collector elements (33 elements in our previous example) must be considerably larger, of course, say 4 times or more larger. If it were desired to maintain the outlet temperature T_o at 60° C even when the inlet temperature were colder than the 50° C just assumed, a thermostat would be placed in the outlet line as shown in Figure 3–5. The thermostat, a temperature-level-actuated valve would throttle the flow, introducing a pressure drop additive to the right hand side of Eq. (3–34) and thus lowering the velocity and mass flow for a given value of ΔP provided according to Eq. (3–33).

With an understanding of the physical principles governing the operation of a solar heat collector, we can now turn to the design procedure.

4. DESIGN STEPS

ESTABLISHING GOALS

Needs. As a first step in the design of a solar collector, the needs are established. For example, for an ordinary domestic water heater, one tries to establish the amount and temperature of hot water needed versus time of day and season of year. One might try to estimate the value to the consumer of each unit of consumption and how much consumption might increase if additional supply were made available at lower cost, or how much consumption might contract if costs were to increase. As an aid to design, one quantifies approximately the value and costs in terms of an appropriate unit, usually a commonly accepted medium of economic exchange, viz., the U.S. dollar.

Constraints. One determines the constraints operating: environmental, social and aesthetic, spatial, and so forth. Chief among the environmental constraints is the amount of solar energy likely to be available as a function of time of day and season of year and the likelihood of interruption of energy supply due to water-droplet and dust-particle clouds.

Costs. One usually sets the goal as meeting the needs, subject to the constraints, at the minimum cost. As discussed briefly in the introduction, costs are those associated with tying up resources, whether the device is used or not, the so-called fixed capital costs, and costs related to the use of the device, such as the cost of electrical power for pumping and/or auxiliary heating. To combine these costs, one normally ascribes an annual cost to the use or recovery of the capital resources and sums it with the other costs to obtain a total annual cost. For example, if \$1000 of capital C must be committed for a solar water heating system, and the sum must be borrowed at $i = 8$ percent interest (or, if available, could be invested elsewhere for that rate), and the capital must be repaid over an expected $n = 20$ year lifetime, then the annual cost c_c associated with use of the capital C is

$$c_c = \left[i + \frac{1}{1+(1+i)+(1+i)^2+(1+i)^3+\ldots+(1+i)^{n-1}} \right] C$$

$$c_c = \frac{i(1+i)^n}{(1+i)^n - 1} C \qquad (4\text{–}1)$$

$$c_c = \frac{(0.08)(1.08)^{20}}{(1.08)^{20} - 1} (1000) = \$101.85$$

To this cost must be added that for taxes, insurance, maintenance and for auxiliary fuel. Uncertain taxes, variable interest rates, and uncertain fuel costs complicate the analysis, needless to say.

DESIGN ALTERNATIVES AND DESIGN VARIABLES

Alternatives. One must conceive of alternatives. To return to a domestic water heater, one may decide upon three main alternatives, (1) a conventional heater, (2) a solar preheater with conventional heater, and (3) a solar heater alone. Within each main alternative a number of subalternatives may exist. The conventional heater may be gas or electric, the solar unit may be any of the many types classified in the introduction, and so forth. Estimating the cost associated with each alternative may bring one to identify (2) above as the best approach. It is then necessary to consider the subalternatives in more detail.

Design Variables. A design variable is really a type of alternative which can be varied at the designer's choice. For example, we have discussed the spacing W between collector flow elements as pictured in Figure 3–1. The thickness b in Figure 3–2 may also be selected, but only discrete sizes are commercially available. The former, W, is thus continuously variable and the latter, b, is variable only in steps.

OPTIMIZATION

The optimization is one of varying the design variables to achieve the design goal, minimum cost-benefit ratio. Consider the effect of the variable W. As it is increased, the fin effectiveness η_f and the collector efficiency η drop, so that a larger total collector size is necessary to meet the design need. However, a lesser amount of conduit, pictured in Figures 3–1 and 3–2, is needed so that the cost of this item is reduced. Let the cost of the conduit per unit length including installation be C_L, say 2 dollars per meter and that of the collector and glazing without tubing C_A be say 100 dollars per square meter. The area needed for a specified $\dot{Q}$ to be collected is given by

$$\dot{Q} = \eta A \bar{q_S}, \qquad A = \dot{Q}/\eta \bar{q_S} \tag{4–2}$$

The capital cost is

$$C = C_A A + C_L N L \tag{4–3}$$

where

$$A = NWL \tag{4–4}$$

Hence

$$C = \frac{\dot{Q}\,C_A}{\bar{q_S}}\left[\frac{1+(R_{LA}L/W)}{\eta}\right] = \frac{\dot{Q}\,C_A}{\bar{q_S}\eta_o}\left[\frac{1+(R_{LA}L/W)}{\eta_f}\right] \tag{4–5}$$

where $\eta = \eta_f\eta_o$ according to Eq. (3–31), and

$$R_{LA} = C_L/C_A L \tag{4–6}$$

As W increases the term R_{LA}/W drops but the coefficient $1/\eta_f$ increases, where η_f is obtained as shown in the development leading up to Eq. (3–29). Clearly an optimum or best value of W exists where C is minimized. The capital cost C is, of course, related to the annual cost by Eq. (4–1).

TABLE 4–1

EFFECT OF CONDUIT SPACING ON COST EFFECTIVENESS

Parameters:

$R_{LA}L = 0.02$ m, $U = 6$ w/m^{2o}C,
$K_w = 380$ w/m°C (copper), $b = 0.25$ mm (0.010 in.)
$W_o = 0.0254$ m, $\delta = 0.12$ mm, $K_a = 0.4$ w/m°C,
$h_i = 170$ w/m^{2o}C, $p_i = 0.072$ mm

Conduit Spacing W (m)	Fin Effectiveness η_f	Cost Effectiveness Parameter $[1+(R_{LA}L/W)]/\eta_f$
0.050	0.9712	1.442
0.075	0.9519	1.331
0.100	0.9279	1.293
0.125	0.9002	1.289 (Best Value)
0.150	0.8695	1.303
0.175	0.8369	1.331
0.200	0.8030	1.370
0.225	0.7687	1.417
0.250	0.7346	1.470

Table 4–1 shows how the term in the braces on the right side of Eq. (4–5) varies with W for the conditions assumed previously for Table 3–1. For these conditions, and for the assumed value of $R_{LA}L = 2/100 = 0.02$ m (note that only the ratio of C_L/C_A affects the table, and not the values themselves), the optimum value of W is found to be approximately 0.125 meters (5 inches). This optimum is not dictated by the physics alone but by the physics of the problem and the economic parameter R_{LA}.

Note that the optimum value of W above is that for an assumed absorber plate thickness b. In order to decide upon the optimum b and W, one needs to know how the plate thickness b affects the cost of the collector C_A, and several tables such as Table 4–1 must be constructed for different values of b, and the optimum value of the product of C_A and the right-hand column must be found as prescribed by minimizing the cost in Eq. (4–5). We may carry out this next step by assuming, for example,

$$C_A = C_{AO}[1 + R_{AB}(b/b_o)] \quad (4\text{–}7)$$

The quantity R_{AB} is the ratio of the cost of the copper absorber of nominal thickness b_o divided by the cost of the cover glass, insulation, manufacture, and installation. In order to carry through to an answer for b and W, a specification of both $R_{LA}L$ and R_{AB}/b_o must be made. Note that under the assumptions made

$$R_{LA} = \frac{R_{LAO}}{1 + R_{AB}b/b_o} \qquad (4\text{–}8)$$

To continue the example, we carry out calculations for $R_{LAO}L = 0.022$ m and $R_{AB}/b_o = 200$ m^{-1}.

TABLE 4–2

EFFECT OF ABSORBER PLATE THICKNESS AND CONDUIT SPACING ON COST EFFECTIVENESS

Parameters:
$R_{LAO} = 0.022$ m, $U = 6$ w/m^{2}°C,
$K_w = 380$ w/m°C (copper), $W_o = 0.0254$ m,
$\delta = 0.12$ mm, $K_a = 0.4$ w/m°C,

Plate Thickness b (mm)	Cost Effectiveness Parameter $\frac{[1 + (R_{AB}b/b_o)]}{\eta_f}\left[1 + \frac{R_{LAO}L}{[1 + (R_{AB}b/b_o)]W}\right]$			
	W = 0.075 mm	W = 0.100 mm	W = 0.125 mm	W = 0.150 mm
0.15	1.398	1.365	1.371	1.402
0.20	1.404	1.365	1.363	1.384
0.25	1.411	1.369	1.362 (Best Value)	1.376
0.30	1.420	1.375	1.365	1.375
0.35	1.429	1.382	1.370	1.377

Table 4–2 shows the results. The optimum value of plate thickness b is found to be 0.25 mm (0.010 inch), and the optimum spacing is 0.125 m (5 inches). Note that the optimum is not particularly sharp. This lack of sharpness allows an experienced designer to exercise intuition to shortcut a completely rigorous optimization procedure, and it reduces the sensitivity of the design specifications to the values of parameters such as $R_{LAO}L$ and R_{AB}/b_o. Even if the values of these parameters are misjudged, the resulting design is not likely to be very far from optimum.

5. CONCLUSION

This work has given the reader an acquaintance with the types of solar collectors which have been developed in the past, a demonstration of how the science of physics can be applied to predict their performance and how the performance is affected by the design features, and a brief description of the procedures used to achieve designs of maximum benefit to users. The types and main components of solar collectors were described as well as measures of their performance. The physics of heat losses from solar collectors and the stratagems taken to reduce those heat losses were reviewed. Included in this review were the fundamentals of spectral selectivity and the concepts of absorptance and transmittance of merit, two measures of the technical value of selective optical components. The honeycomb collector was briefly described. Attention was then directed to the heat transfer from the solar absorber to the working fluid that picks up and carries the solar heat energy to its intended use. The concepts of exchanger effectiveness and number of transfer units were employed. So was the concept of fin effectiveness, needed to determine one parameter affecting the heat exchanger. The presentation turned finally to a brief discussion of design optimization, and the rather prosaic problem of optimizing the absorber plate thickness and collection tube spacing of a flat plate solar water heater was offered as a concrete example of the optimization procedure.

It is hoped that the reader will be stimulated into undertaking the design of a solar heat collector for his or her own use, or will evaluate those commercially available. The point has been made that solar energy is not "free" in that valuable resources must be committed to collect it and that these resources ought not to be wasted. When solar heaters are used in combination with conventional auxiliary ones for maximum availability of energy, the designer will discover that somewhat undersized solar collectors make the best use of available resources. By designing the unit to accommodate minimum needs, the designer ensures that all or nearly all of the solar energy collected is useful, and the resources committed to collecting it are not therefore wasted.

APPENDIX

Sample Calculation of Total Solar Absorptance and Total Emittance from Spectral Data

SOLAR ABSORPTANCE

To illustrate concretely the use of the information in Fig. 2-3, spectral data fo commercial-finish, galvanized iron in Fig. 2-7 are reduced to a value of total solar absorptance. Equation (2-14) is approximated by a numerical summation using the simple trapezoidal rule.

$$\alpha = \int_0^1 \alpha(\lambda) df(\lambda) \doteq \sum_i \alpha(\lambda_i)\Delta_i f \qquad \text{(A-1)}$$

First a table of $\Delta_i f$ is arbitrarily created, and from that table centrally located values of $f(\lambda_i)$ are found. Figure 2-3 then allows λ_i values to be read, and Fig. 2-7 gives $\alpha(\lambda_i)$ from $1-\rho(\lambda_i)$ for those values. The sum in Eq. (A-1) is then obtained. Table A-1 shows a 16-point numerical summation leading to a value of $\lambda = 0.58$.

TABLE A-1

NUMERICAL INTEGRATION FOR SOLAR ABSORPTANCE

i	$\Delta_i f$	f_{lower}	f_{upper}	f_{mean}	λ_i	α_i	$\alpha_i \Delta_i f$
1	.02	0	.02	.01	.36	.57*	.0114
2	.02	.02	.04	.03	.39	.57*	.0114
3	.02	.04	.06	.05	.42	.57*	.0114
4	.04	.06	.10	.08	.44	.57*	.0228
5	.10	.10	.20	.15	.48	.57	.0570
6	.10	.20	.30	.25	.55	.57	.0570
7	.10	.30	.40	.35	.61	.58	.0580
8	.10	.40	.50	.45	.68	.62	.0620
9	.10	.50	.60	.55	.76	.64	.0640
10	.10	.60	.70	.65	.85	.68	.0680
11	.10	.70	.80	.75	1.05	.74	.0740
12	.10	.80	.90	.85	1.3	.48	.0480
13	.04	.90	.94	.92	1.6	.36	.0144
14	.02	.94	.96	.95	1.8	.34	.0068
15	.02	.96	.98	.97	2.1	.30	.0060
16	.02	.98	1.00	.99	3.0	.25	.0050
							$\sum_i = 0.577$
							$\alpha = 0.58$

*Extrapolated

The reader may repeat the calculation for a corrugated absorber plate where two solar reflections occur before escape. Then $\alpha(\lambda_i) = 1 - \rho^2(\lambda_i)$. Also the reader may experiment with other numerical integration schemes. The reader should note that some extrapolation of the spectral data was necessary to carry out the calculation, and numerical sensitivity to such extrapolation can be evaluated as well.

INTERNAL TOTAL EMITTANCE

Now consider a 100 °C absorber plate exchanging radiation with a 40 °C cover. Suppose the internal total emittance is desired for the mean temperature of 70 °C. Equation (2-21) is approximated as was done with Eq. (2-14).

$$\epsilon(T,T_C) = \int_0^1 \alpha(\lambda,T)df_i(\lambda T,T_C/T) \doteq \sum_i \alpha(\lambda_i,T)\Delta_i f_i \qquad \text{(A-2)}$$

Again a table of $\Delta_i f$ is arbitrarily created, and values of $f_i(\lambda_i T,T_C/T)$ from Fig. 2-5 are used to find $\lambda_i T$ values. For the mean value of T = 70 °C = 343.2 °K, the values of λ_i

TABLE A-2

NUMERICAL INTEGRATION FOR INTERNAL-TOTAL EMITTANCE

i	$\Delta_i f_i$	f_{mean}	$\lambda_i T$ μm ° K	λ_i μm	$\alpha_i(\lambda_i,T_O)$	$\alpha_i(\lambda_i,T_S)$	$\alpha_i \Delta_i f_i$
1	.02	.01	1250	3.6	.20	.22	.0044
2	.02	.03	1500	4.4	.17	.19	.0038
3	.02	.05	1600	4.7	.17	.19	.0038
4	.04	.08	1750	5.1	.16	.18	.0072
5	.10	.15	2000	5.8	.14	.16	.0160
6	.10	.25	2300	6.7	.14	.16	.0160
7	.10	.35	2600	7.6	.13	.15	.0150
8	.10	.45	3000	8.7	.13	.15	.0150
9	.10	.55	3400	9.9	.13	.15	.0150
10	.10	.65	4000	11.7	.12	.13	.0130
11	.10	.75	4700	13.7	.12	.13	.0130
12	.10	.85	5800	16.9	.11*	.12	.0120
13	.04	.92	7300	21.3	.10*	.11	.0044
14	.02	.95	8800	25.6	.09*	.10	.0020
15	.02	.97	11000	32.1	.08*	.09	.0018
16	.02	.99	16000	46.6	.07*	.08	.0016

*Extrapolated

$\sum_i = .144$

are found, and, for each value of λ_i, absorptance $\alpha(\lambda_i,T_S) = 1-\rho(\lambda_i,T_S)$ is taken from Fig. 2-7. Note that the actual surface temperature pertains to the spectral property data.

One problem remains. The values of $\rho(\lambda_i,T_S)$ plotted in Fig. 2-7 come from data taken at a surface temperature of $T_O = 300\,°K$. It is necessary to assume a temperature variation for the infrared spectral absorptance. A variation of $(T_S/T_O)^n$ with $n = 1/2$ is assumed here for illustration. The reader may repeat the calculation for say $n = 1$ and for other mean temperatures to see the sensitivity of the result to the assumed values. The surface temperature dependence of spectral absorptance in the solar region was neglected, because it is usually small.

Table A-2 gives the sample calculation. A total value of $\epsilon = 0.14$ is obtained. One final note is that the data in Fig. 2-7 pertain to a near-normal angle of incidence. The calculated total absorptance value is thus appropriate for such incidence. However, the emittance value should be an average one as discussed briefly in the section on directional selectivity. For a very low-emittance metal, the hemispherical emittance is up to 4/3 of the normal emittance. A value approximately 10 to 15 percent larger may be appropriate for the galvanized surface used here for illustration. Thus one would take ϵ_H to be approximately 0.16.

More precise calculations would account for the spectral and directional variations of both the absorber plate and coverglass properties for each component of polarization. The method of using Figs. 2-3 and 2-5 as illustrated here remains applicable.

SUGGESTED READING

I. HEAT TRANSFER

A. *Natural Convection*

The pioneer paper is "Radiation, Convection and Conduction Coefficients in Solar Collectors," by H. Tabor, *Bulletin of the Research Council of Israel, Volume 6C,* Number 3, pages 155–176, August 1958. A paper reviewing more recent findings is "Natural Convection in Enclosed Spaces—A Review of Application to Solar Energy Collection," by H. Buchberg, I. Catton, and D. K. Edwards, *Journal of Heat Transfer, Volume 98,* pages 182–188, May 1976. Heat transfer across the space between absorber plate and cover glass is reviewed, with and without honeycomb material being present.

B. *Radiation*

The textbook *Thermal Radiation Heat Transfer,* by R. Siegel and J. R. Howell, McGraw-Hill Book Co., New York, 1972, gives a thorough presentation of the general subject of radiation heat transfer. The older book, *Radiation Heat Transfer for Spacecraft and Solar Power Plant Design,* by F. Kreith, International Textbook Co., Scranton, Pa., 1962, is directed more specifically to those aspects of the subject of concern in solar energy applications. A book directly concerned with earthbound applications in heating and cooling is *Thermal Environmental Engineering,* by J. L. Threlkeld, Prentice-Hall, Inc., Englewood Cliffs, New Jersey, 1962. See also the topic Spectral Selectivity in this list.

II. SOLAR COLLECTORS

A. *Concentrating Systems, Low Concentration*

A low-concentration system demonstrated at the United Nations Conference on New Sources of Energy, Rome, Italy, 1961, is described in "Low-Cost Focusing Collector for Solar Power Units," by H. Tabor and H. Zeimer, *Solar Energy, Volume 6,* Number 2, pages 55–59, 1962. Two other works are "Mirror Boosters for Solar Collectors," by H. Tabor, *Solar Energy, Volume 10,* Number 3, pages 111–118, 1966, and "Moderately Concentrating Flat-Plate Solar Energy Collectors," by R. B. Bannerot and J. R. Howell, American Society of Mechanical Engineers, Paper number 75-HT-S4, 1975.

B. *Concentrating Systems, High Concentration*

Three high-concentration, field-of-mirrors systems are described in "Pilot Plants of Solar Steam Generation Stations," by G. Francia, *Solar Energy, Volume 12,* pages 51–64, 1968. Recent thinking in the U.S. in regard to high-concentration systems is represented by "Central Collector Solar Energy Receivers," by A. Sobin, W. Wagner, and C. R. Easton, *Solar Energy, Volume 18,* pages 21–30, 1976, and "Solar Thermal Power System Based on Optical Transmission," by L. L. Vant-Hull, ibid., pages 31–39, 1976.

C. *Flat Plate Collectors*

Two pioneer papers are "The Performance of Flat-Plate Solar-Heat Collectors," by H. C. Hottel and B. B. Woertz, *Transactions of the American Society of Mechanical Engineers, Volume 64,* pages 91–104, 1942, and "Evaluation of Flat-Plate Solar Collector Performance," by H. C. Hottel and A. Whillier, *Transactions of the Conference on the Use of Solar Energy, Volume 3,* (*Thermal Processes*), Part 2, pages 74ff, 1955. Another one is "The Derivations of Several Plate Efficiency Factors Useful in the Design of Flat-Plate Solar Heat Collectors," by R. W. Bliss, Jr., *Solar Energy, Volume 3,* Number 4, pages 55–64, 1959. Two papers from a large literature on water heaters are "The Performance of Solar Water Heaters with Natural Circulation," by D. J. Close, *Solar Energy, Volume 6,* Number 1, pages 33–40, 1962, and "Effect of Materials and Construction Details on the Thermal Performance of Solar Water Heaters," by A. Whillier and G. Saluja, *Solar Energy, Volume 9,* pages 21–26, 1965. Some other representative works are "A Rational Procedure for Predicting the Long-Term Average Performance of Flat-Plate Solar-Energy Collectors," by B. Y. H. Liu and R. C. Jordan, *Solar Energy, Volume 7,* Number 2, pages 53–74, 1963, and "A Design Approach for Solar Processes," by D. J. Close, *Solar Energy, Volume 11,* Number 2, pages 112–122, 1967.

D. *Honeycombs*

A selectively-transparent glass honeycomb system with tracking and concentration was described in the paper, "A New Collector of Solar Radiant Energy—Theory and Experimental Verification," by G. Francia, U.N. Conference on New Sources of Energy, Rome, Italy, 1961, Paper E/CONF. 35/S/71.

Tests of a fixed rectangular honeycomb are reported in "Performance Characteristics of Rectangular Honeycomb Solar-Thermal Converters," by H. Buchberg, O. A. Lalude, and D. K. Edwards, *Solar Energy, Volume 13,* pages 193–221, 1971. Design of glass honeycomb systems is treated in "Design Considerations for Solar Collectors with Cylindrical Glass Honeycomb," by H. Buchberg and D. K. Edwards, *Solar Energy, Volume 18,* pages 193–203, 1976. Plastic honeycombs are described in "Use of Lexan and Kapton Honeycombs to Increase Solar Collector Efficiency," by K. N. Marshall and R. K. Wedel, 16th National Heat Transfer Conference, August 8–11, 1976, St. Louis, Missouri, paper number A.I.Ch.E.–55, available from the Engineering Societies Library, United Engineering Center, 345 East 47th Street, New York. Other papers of interest are "Natural

Convection in Enclosed Spaces—A Review of Application to Solar Energy Collection," by H. Buchberg, I. Catton, and D. K. Edwards, *Journal of Heat Transfer, Volume 98,* pages 182–188, May 1976; "Honeycomb Devices in Flat-Plate Solar Collectors," by K. G. T. Hollands, *Solar Energy, Volume 9,* Number 3, pages 159–164, 1965; and "Performance Measurements of a Cylindrical Glass Honeycomb Solar Collector Compared with Predictions," by J. C. McMurrin, N. A. Djordjevic, and H. Buchberg, American Society of Mechanical Engineers Winter Annual Meeting, 1976, Paper Number 76-WA/Sol-3, accepted for publication (in slightly revised form) in the *Journal of Heat Transfer.*

E. *Vee-Groove Collectors*

The paper "Directional Selectivity, Emittance, and Absorptance Properties of Vee Corrugated Specular Surfaces," by K. G. T. Hollands, *Solar Energy, Volume 7,* Number 3, pages 108–116, 1963, presented the concept of the vee-groove absorber and "Solar Air Heaters for Low and ModerateTemperature Applications," by D. J. Close, ibid., pages 117–124, described their application as air heaters.

F. *Solar Ponds*

Pioneer works are "The Physics of the Solar Pond," by H. Weinberg, *Solar Energy, Volume 8,* Number 2, pages 45–56, 1964, and "The Solar Pond Project," by H. Tabor and R. Matz, *Solar Energy, Volume 9,* Number 4, pages 177–182, 1965.

G. *Stills*

Knowledge gained from operating water-desalting stills up to 465 m^2 in size is summarized in "Design Philosophy and Operating Experience for Australian Solar Stills," by P. I. Cooper and W. R. W. Read, *Solar Energy, Volume 16,* pages 1–8, 1974.

III. SPECTRAL SELECTIVITY

The advantages to be gained from spectrally selective absorbers were propounded in two papers at the 1955 Conference on the Use of Solar Energy in Tucson, Arizona, "Selective Radiation. I. Wavelength Discrimination. II. Wavefront Discrimination," by H. Tabor, Transactions of the Conference on the Use of Solar Energy, The Scientific Basis, Tucson, Arizona, October 31–November 1, 1955, Ch. 3 of Vol. II, Thermal Processes, Part 1, Section A, Flat Plate Collectors, Univ. of Arizona Press, pages 1–39, 1955, and "Selective Spectral Characteristics as an Important Factor in the Efficiency of Solar Collectors," by J. T. Gier and R. V. Dunkle, ibid., pages 41–56. Several optical-system concepts and some spectral data were reported in "Spectral and Directional Thermal Radiation Characteristics of Selective Surfaces for Solar Collectors," by D. K.

Edwards, J. T. Gier, K. E. Nelson, and R. D. Roddick, U.N. Conference on New Sources of Energy, Rome, Italy, August 1961, Paper E/CONF. 35/S/43. (Also *Solar Energy, Volume 6,* Number 1, pages 1–8, 1962). The basis for ready comprehension of the behavior of selective surfaces was laid in "Thermal Radiation Tables and Applications," by R. V. Dunkle, *Transactions of the American Society of Mechanical Engineers, Volume 76*, pages 549–552, 1954. Recent investigations include "Solar Absorptance and Emittance Properties of Several Solar Coatings," by R. B. Pettit and R. R. Sowell, and "Spectral Reflectance Properties of Plated Zinc for Use as a Solar Selective Surface," by G. E. McDonald, reported at the 1975 International Solar Energy Congress, July 28–August 1, 1975, University of California, Los Angeles.